Institutions and Collective Action

A publication of
the Center for Self-Governance

Institutions and Collective Action

Self-Governance in Irrigation

Shui Yan Tang

ICS PRESS
Institute for Contemporary Studies
San Francisco, California

This book is a publication of the Center for Self-Governance, dedicated to the study of self-governing institutions. The Center is affiliated with the Institute for Contemporary Studies, a nonpartisan, nonprofit, public policy research organization. The analyses, conclusions, and opinions expressed in ICS Press publications are those of the authors and not necessarily those of the Institute, or of its officers, directors, or others associated with, or funding, its work.

Inquiries, book orders, and catalog requests should be addressed to ICS Press, 243 Kearny Street, San Francisco, CA 94108. (415) 981-5353. Fax (415) 986-4878. To order, call toll free in the contiguous United States: **(800) 326-0263**. Distributed to the trade by National Book Network, Lanham, Maryland.

Library of Congress Cataloging-in-Publication Data

Tang, Shui Yan.
    Institutions and collective action: self-governance in irrigation
/ by Shui Yan Tang.
        p.   cm.
    Includes bibliographical references (p.   )
    ISBN 1-55815-179-6 (paper) $9.95
    1. Irrigation—Government policy—Developing countries—Citizen participation.   2. Irrigation—Government policy—Developing countries—Citizen participation—Case studies.   I. Title.
    HD1741.D44T36   1992
    333.91′3′091724—dc20   91–42624
                                                                        CIP

To the memory of my parents,
Pui-Yin Tang and Oei-Yu Luk

CONTENTS

L I S T   O F   T A B L E S

LIST OF FIGURES

FOREWORD

Our natural resources are in jeopardy. Conservation is not enough; to forestall disaster, we must change the way we think about these resources. It is imperative that we reconsider not only how they should be used but also how their use should be governed.

When we think of how water, forests, fisheries, grazing land, and other natural resources are best managed, the first idea that comes to most of us is "by the government." Traditionally, government bureaucracies on the national or state level have controlled how these resources are used, setting the rules for who could have access to how much, how often, and at what price. The government, so the reasoning went, was in the best position to administer the resource fairly, establishing regulations and enforcing compliance among the users.

But is this true? Is direct oversight by a government bureaucracy the best answer to natural resource management? Or can the resource users—who, after all, have the greatest stake in the resource's management—be just as effective in governing themselves? Are local, self-governing organizations, using collective action, more apt to maintain the economic viability of the resource than a bureaucratic institution?

This new study by Shui Yan Tang helps refocus our thoughts on these critical questions. Forty-seven case studies of irrigation systems around the world provide a wealth of information about governing organizations that work and those that don't. Tang's careful analysis of the data helps us understand the wide diversity of experiences with the relationship between institutions and collective action in resource management. The results suggest that systems that include elements of self-governance—in which rules are established and compliance is enforced by users, rather than by an outside bureaucracy—can work well.

As Tang points out, how well each system works depends on the physical, social, and institutional factors operating in each case. These factors interact to create the unique set of needs and problems that must be handled by each governing institution. He is clear in presenting both the potential and the limitations of self-governance in irrigation systems, and he realistically discusses how bureaucratic and local self-governing organizations can work together in effective partnership.

Understanding self-governance and the institutions through which it can be realized—not just in natural resource management but in every aspect of our lives—can make a vital difference in how we live. It holds the key to the quality of our future. We believe that the information in this book, and in others produced by the Center for Self-Governance, will contribute to our application of self-governing principles to an ever-widening range of endeavors.

Robert B. Hawkins, Jr., President
Institute for Contemporary Studies

P R E F A C E

This book is built on the theoretical and empirical work of many individuals. Institutional analysis has been a major research interest for scholars associated with the Workshop in Political Theory and Policy Analysis at Indiana University, where I was a graduate student for five years. Scholars in the workshop are concerned with the effects of rule-ordered relationships on governance and development in human societies. In this book, I apply a mode of analysis developed by these and other scholars to examine how institutional arrangements affect collective action in irrigation systems.

The empirical part of this book is based on existing case studies in the literature. The Common-Pool Resource (CPR) Project in the Workshop in Political Theory and Policy Analysis has been collecting case studies on irrigation systems and other common-pool resources. Colleagues in the project have also developed a series of in-depth coding forms and computer software to code and analyze information contained in these case studies. I drew heavily on these research tools and materials in the empirical analysis for this book.

In the course of preparing this book, I have received generous help and support from many individuals and organizations. My deepest appreciation goes to Elinor Ostrom, who has been my mentor and both the intellectual and organizational leader of the CPR Project. Lin has offered me so many ideas and insights that no number of citations in the book can adequately acknowledge my intellectual debt to her.

I benefited tremendously from the numerous hours of discussions with other colleagues associated with the CPR Project: Arun Agrawal, Libby Anderson, William Blomquist, Laura Celis, Jamal Choudhry, Roy Gardner, Edella Schlager, James Walker, and James Wunsch. They, Julie England, and Fenton Martin have also contributed to the collection of case materials and the development of other research instruments on which the research of this book is built. Readers of various versions and parts of the manuscript included Wesley Bjur, Gerald Caiden, Robert Hunt, Larry Kiser, Elinor Ostrom, Vincent Ostrom, Roger Parks, James L. Perry, R. K. Sampath, Edella Schlager, and Michael Woolley. They have provided helpful comments on the manuscript. Needless to say, I am responsible for any remaining faults.

The research in this book was supported by grants from the National Science Foundation (SES-8619498) and the U.S. Agency for International Development (DHR-1066-GSS-6042). Another grant from the U.S. Agency for International Development (DHR-5446-Z-00-7033-00) supported my trip to Nepal in March 1989, where I benefited from discussions with scholars and officials about irrigation systems and policies in Nepal. The School of Public Administration at the University of Southern California has provided a pleasant working environment in which to finish the manuscript.

Finally, I would like to thank Robert B. Hawkins, Jr., of the Institute for Contemporary Studies, for encouraging me to prepare the manuscript for publication. Editorial support from Patty Dalecki, Janet Mowery, and Barbara Kendrick is also gratefully acknowledged.

Institutions and Collective Action

CHAPTER ONE

Local, Self-Governing Organizations and Irrigation Development

What institutional arrangements can effectively help in the governance of such natural resources as inshore fisheries, grazing land, and water systems? Is direct management by national governments the most effective way of governing these resources? In what circumstances can local, self-governing organizations effectively ensure the long-term economic viability of these resources? What factors affect the performance of these self-governing organizations? In what ways does government intervention affect the functioning of these organizations? This book is an effort to provide answers to some of these questions based on the cumulative work of scholars who study the performance of diverse institutional arrangements and on an empirical analysis of the governance arrangements for one type of resource: irrigation systems.

Irrigated agriculture is the major means of subsistence for people in many parts of the world. Improving the performance of irrigation systems has become more important as the total irrigated areas of the world nearly tripled between 1950 and 1986, amounting to around 250 million hectares in 1986 (Postel 1990, 40). Many developing countries and international agencies have invested enormous resources in the development of irrigation systems. As a result of various institutional problems, however, many of these systems have failed to meet operational targets or have deteriorated rapidly soon after they were constructed. Improving the operation

1

and maintenance of these irrigation systems is an important means of increasing agricultural production in many developing countries (Hayami and Ruttan 1985).

Operating and maintaining an irrigation system requires coordination among many farmers. Collective-action problems arise easily when each farmer has the incentive to use more water and invest less in the system. These problems often result in poor maintenance as well as conflicts and anarchy in water allocation. Their solution requires institutional arrangements to provide a structure of rules that enable participants to sustain credible commitments and long-term productive relationships with one another.

In many indigenous, community irrigation systems, farmers have developed a wide diversity of rules to specify rights and responsibilities among themselves. Farmers enforce these rules by themselves without involving external authorities. These rules enable farmers to cooperate in the operation and maintenance of their irrigation systems. In Thulo Kulo, a farmer-owned irrigation system in Nepal, for example, users are joint owners of all aspects of the water delivery system (Martin and Yoder 1986). Different users are entitled to various amounts of water depending on the number of shares they own. Each shareholder is required to contribute a corresponding amount of labor to maintain the system.

Other farmers have developed multilayered organizations to govern their irrigation systems, such as the *zanjera* irrigation systems in the Philippines (Siy 1982; Coward 1979). In each *zanjera*, rules are established to determine how to select officials and how to divide construction and maintenance duties among members. Several *zanjeras* sharing a water source may form a larger federation. The Bacarra-Vintar Federation of *Zanjeras*, for example, is a federation of nine *zanjeras* (Siy 1982). The heads of all the component *zanjeras* form the Board of Directors for the federation. While each *zanjera* is responsible for its own internal affairs and has no monetary obligations to the federation, the federation has the authority to assign responsibilities to individual *zanjeras* regarding the annual maintenance of the dam that diverts water from a nearby river to all *zanjeras*. The federation also acts as a corporation that has secured water rights from the Philippine government.

In large-scale, government-run irrigation systems, farmers' organizations can also play an active role in providing operation and maintenance at the watercourse level. For instance, farmers in large bureaucratic irrigation systems in India and Pakistan sometimes form their own organizations within their watercourses. In one of these organizations, described in depth by Wade (1988a), farmers select their own officials and enforce their own water allocation rules without the explicit endorsement of government officials. The organization helps to resolve conflict among farmers and to ensure that the flow of water to their watercourse is not obstructed by any upstream communities.

A common feature shared by all these organizations is self-governance. These organizations are not created by government. Instead, farmers themselves develop rules that assign rights and responsibilities among themselves. They are responsible for enforcing the rules they create and for resolving disputes among themselves. Although these organizations may benefit from support offered by external authorities, their existence and functioning do not entirely depend on these authorities.

These examples provide invaluable information about the potential and the limitations of self-governing organizations for solving collective-action problems in irrigation systems. Learning from these examples requires a systematic strategy. It is necessary first to identify the types of collective-action situations facing participants in the resource; then to examine how various physical, social, and institutional factors affect the relationships among the participants in those situations; and finally to begin to understand those circumstances that allow farmers to solve their collective-action problems through self-organization and those that would benefit from government intervention.

In this chapter, I first describe some typical collective-action situations faced by irrigators. I then examine the role of local, self-governing organizations in resolving collective-action problems in irrigation. I conclude the chapter by outlining the plan of this book.

Collective-Action Problems in Irrigation Systems

The physical attributes of a resource affect the relationships among the users and potential users. Two independent attributes—the feasibility of exclusion and subtractibility—are frequently used to classify resources or goods (see V. Ostrom and E. Ostrom 1977;

Gardner, Ostrom, and Walker 1990). *Exclusion* occurs when potential users can be denied goods unless they meet certain criteria. A good is *subtractive* when one person's use of it prevents its use by others. If these two attributes are arrayed in a simple matrix, four types of goods can be identified: common-pool resources, public goods, private goods, and toll goods (see Table 1.1).

As common-pool resources, irrigation systems are characterized by two features. First, it is costly (but not necessarily impossible) to exclude potential beneficiaries from using an irrigation system once it has been constructed. In most cases, the cost of exclusion is due to the size of the water delivery facilities and the flow nature of water. Second, the flow of water available at any one time in an irrigation system is limited. The use of water by one individual therefore subtracts from the amount available to others.

Public goods, such as national defense, are similar to common-pool resources in that it is difficult to exclude potential beneficiaries from enjoying them once they have been provided. They differ, however, from common-pool resources in that once a public good is available, one individual's enjoyment of it does not subtract from the amount available to others. The most important collective-action problem concerning public goods is the organization of provision. Individuals have little incentive to contribute to the provision of public goods because it is difficult to exclude potential beneficiaries from enjoying them. Once the provision prob-

---

TABLE 1.1   Classification of Resources and Goods

	Subtractive consumption	Nonsubtractive consumption
Costly to exclude	Common-pool resources (e.g., irrigation systems)	Public goods (e.g., national defense)
Not costly to exclude	Private goods (e.g., bread)	Toll goods (e.g., cable TV)

SOURCE: Adapted from V. Ostrom and E. Ostrom, "Public Goods and Public Choices," in *Alternatives for Delivering Public Services: Toward Improved Performance*, ed. E. S. Savas (Boulder, Colo.: Westview Press, 1977), 7–49.

lem is solved, however, the allocation of public goods is not a problem since they can be simultaneously enjoyed by many individuals without significant depletion.

Like public goods, common-pool resources involve the *provision* problem because it is costly to exclude potential beneficiaries. Common-pool resources have another important problem: It is necessary to *regulate the use of the limited amount of flow units available* in order to ensure productive and equitable uses. In irrigation systems, water allocation and provisions are two major sources of collective-action problems.

Water allocation is a major source of conflict in irrigation. When the amount of water is not sufficient to satisfy everyone's cultivation needs simultaneously, farmers face the prospect of decreases in crop yields or even losses of entire crops. It is not uncommon to see this conflict develop into bloodshed or even murder (Maass and Anderson 1986). Whenever the demand for water exceeds the supply, some kind of allocation process needs to be specified by a set of rules. The allocation of water may be based on the number of water shares held, the amounts of farmland cultivated, or a wide variety of other criteria. Regardless of the bases of allocation, the need to adopt allocation rules means that some farmers will obtain less water than they desire. As the supply of water decreases, the temptation for individual farmers or groups of farmers to break the rules increases.

The situation farmers face frequently resembles a "prisoner's dilemma" game (see Wade 1988b, 490). For all farmers, the alternatives are either to get more water or to refrain from getting more water than they are entitled to use (Weissing and Ostrom 1991). If the flow of water is equally accessible to all farmers, each farmer's preferred ordering of outcomes is as follows: (1) he does not refrain while others refrain; (2) everybody, including himself, refrains; (3) nobody refrains; (4) he refrains while others do not. If outcome 1 occurs, the farmer is a "free rider"; if outcome 4 occurs, he is a "sucker." If every farmer tries to be a free rider and avoids being a sucker, the collective outcome is 3, in which nobody refrains. This outcome is inferior to outcome 2, in which everybody refrains. The outcome when no one refrains is an example of what Garrett Hardin (1968) calls "the tragedy of the commons" or what others call "the commons dilemma" (Gardner, Ostrom, and Walker 1990).

Farmers frequently have to invest substantial resources to construct and maintain facilities such as dams, canals, and pumps that are essential for diverting and transporting water. In some indigenous irrigation systems in Nepal and the Philippines, for example, each farmer devotes about one month of hard physical work every year to repairing and maintaining the water delivery system (Martin and Yoder 1986; Siy 1982). Irrigators have to discipline themselves and their animals to keep an irrigation system in good condition: Pushing a valve or a gate too hard may break an important part of the water control facility; letting heavy animals walk across a canal may damage it. In some instances, individual irrigators may get more water to their fields by making cuts at ditches. To restrain themselves, they must forgo immediate benefits by refraining from activities that harm irrigation facilities. All these efforts are investments by irrigators in the irrigation system.

As in water allocation, farmers face the "commons dilemma" in their investments in irrigation facilities. Because it is often difficult to exclude other irrigators from enjoying the benefits of an operating system, individuals have incentives to refrain from investing, hoping to benefit from others' contributions. If everyone acts likewise, there will be "underinvestment" in constructing and maintaining irrigation facilities. Again, as in water allocation, even if all farmers have promised to contribute, some may still be tempted to withhold their contributions, hoping others will do the job for them. This is, of course, another manifestation of the "free-rider" problem associated with the difficulty of excluding beneficiaries from a collective good (Olson 1965).

The metaphors of the commons dilemma and free riding underscore a potential irony in the human world: Rational human beings may fail to cooperate to produce mutually beneficial outcomes. The tragedy of the commons arises easily when many individuals are involved and when they have difficulty in communicating and enforcing agreements among themselves. In such situations, certain individuals believe that their actions will *not* have a major impact on their own returns but will have a perceptible impact on others' actions. Even if these individuals were to follow a cooperative strategy, it would make little difference to the strategies adopted by other participants.

The tragedy of the commons is inevitable, however, only when participants have *no* control over the structure of the situation they face (E. Ostrom 1988). When an identifiable group of individuals is involved and all participants are aware of the effects of their actions on the actions of others, individuals may be able to develop institutional arrangements that change the structure of their situation. Institutional arrangements that effectively monitor and impose sanctions on rule breakers create incentives for individuals to cooperate in water allocation and investment.

Many actual irrigation situations are more complex than the simplified version just presented. In many irrigation systems, for example, irrigators cultivating crops in the head portion of a canal have a more secure supply of water than those in the tail portion whether or not any allocation rule is in place. Headenders may initially be less motivated to find institutional solutions to water allocation problems than tailenders. In some situations, however, the need for coordination in investment may induce headenders to cooperate with tailenders in devising rules that protect the interests of both parties. The contributions of tailenders in constructing and maintaining water diversion works, such as dams and pumping devices, could lessen considerably the burdens of headenders. In order to solicit such contributions, headenders may have to make concessions to tailenders regarding water allocation.

Thus the allocation of water and the decision to invest are two major sources of collective-action problems in irrigation systems. Some of these problems have incentive structures similar to those involved in the prisoner's dilemma. Nevertheless, when additional factors such as differences in location are considered, the kinds of problems facing many farmers are even more complex than those of the symmetrical prisoner's dilemma game presented in introductory textbooks. Whether presented as simple or complex problems, water allocation and investment problems exist in all irrigation systems. These problems have to be solved by some form of institutional arrangement, or substantial conflict and inefficient outcomes will result.

The Role of Local, Self-Governing Organizations in Irrigation Development

Bureaucratic governance is often proposed as an effective means of solving collective-action problems in common-pool resources.

The experiences of large bureaucratic irrigation systems in many developing countries, however, show that bureaucratic governance is not a panacea for the commons dilemma. In the words of Wade:

> Many theorists of Prisoner's Dilemma have concluded that the socially desirable outcome of restrained access by all can be sustained only by an external authority imposing penalties against rule violation. If so, the conclusion is a counsel of despair. In the reality of most Third World countries, legal mechanisms and the authority of government are simply not powerful enough to make a sufficiently plausible threat across myriad micro situations. (1988b, 490)

Although massive resources have been invested in constructing large public irrigation systems in developing countries, many of these systems have deteriorated rapidly as a result of faulty design and construction, insufficient maintenance, and ineffective operation. The performance of many of these systems has fallen seriously short of expectations: Area irrigated, yield increase, and efficiency in water use are usually much less than initially projected (Repetto 1986, 3–4).

This dismal scene in many bureaucratic irrigation systems contrasts with the experiences of the community organizations mentioned earlier. In these organizations, farmers are able to construct, maintain, and operate their own irrigation facilities effectively. Although large-scale public irrigation systems and small-scale community irrigation systems are technologically very different, development agencies and governments in developing countries have devoted more attention to the potential contributions of community organizations in irrigation development.

These governments and agencies have begun to emphasize the organization of farmers at the watercourse level in large-scale bureaucratic irrigation systems. Unfortunately, these governments and agencies often fail to assess correctly the potential and limitations of local, self-governing organizations. They often mandate the creation of "farmer organizations" as central directives without considering farmers' incentives and capabilities. Douglas Merrey has documented such a problem in Pakistan:

> New legislation was adopted in each province, ostensibly enabling the establishment of water users associations but in fact strengthen-

ing the power of the state over the watercourse. Farmers are obliged to carry out maintenance themselves or repay the costs if the government does it for them.... Government officials delegated to project areas by the Provincial Government retain control of water and other resources and continue to respond to directives from the provincial capital rather than to the demands of local farmers. All of these activities are directed at trying to impose state wishes at the local level, but they do not address the fundamental organizational issues in Pakistan's irrigation management structure. (Merrey and Wolf 1986, 23–24)

Merrey indicates that after these kinds of measures have been implemented, serious allocation and maintenance problems persist in many watercourses.

Robert Hunt (1989) has also cautioned against using the crude analogy of community irrigation systems to water users' associations within bureaucratic irrigation systems. He argues that community irrigation systems are self-contained entities sustained by a set of mutually supportive institutional arrangements, whereas water users' associations are units within a larger bureaucratic environment, the proper functioning of which requires careful communication and coordination across different organizational levels. The participation of farmers in bureaucratic irrigation systems will succeed only if both the organizational problems of the bureaucratic machinery and the structure of incentives facing irrigators are corrected. If the supply of water to a watercourse is highly unpredictable and depends entirely on the arbitrary decisions of officials operating at the system level, it is hard to expect farmers to organize among themselves to undertake operation and maintenance at the watercourse level. Based on experiences in south Asia, Robert Chambers also argues that "main system management to ensure an adequate, convenient, predictable and timely water supply to the outlet is a precondition for farmers' willingness" to undertake activities "such as construction and maintenance of field channels and land-sharing for irrigation" (Chambers 1988, 90).

Farmers may be involved in decision making and organization in a bureaucratic irrigation system in numerous ways. In some cases, they may be responsible for making operational decisions at the tertiary canal or watercourse level only. In other cases, they may participate in making operational and collective decisions at both the watercourse and system level. Depending on the extent to which farmers are involved in an irrigation system and the institutional arrangements that structure their participation, the system may perform differently.

Although involving farmers in the operation and maintenance of bureaucratic irrigation systems is a major policy concern in countries such as India, Pakistan, and Sri Lanka, government assistance to community irrigation systems is an important policy area in countries such as Nepal and the Philippines where many irrigation systems are owned and managed by farmers. Some observers argue that the performance of many community irrigation systems could be improved by various kinds of external assistance, such as the provision of financial or material resources, to strengthen or extend existing diversion structures and water delivery channels (Pradhan 1988).

In countries such as Nepal and the Philippines, specialized government programs and agencies have been established to assist farmers in community irrigation systems. In Nepal, for example, various public and development agencies such as the Department of Irrigation, the Department of Agriculture, and the Agricultural Development Bank have recently been involved in assisting community irrigation systems. These government interventions have met with mixed results. In some cases, farmers benefited from the assistance; in other cases, the assistance created additional conflict and problems among farmers.

Potential pitfalls exist for government intervention in community irrigation systems. A common pitfall is the failure to recognize the conditions that make local, self-governing organizations viable. In the "Proposal for Decentralization Program Support in Nepal," Elinor Ostrom writes:

> Indigenous institutions rely upon shared understandings of rights and duties to enforce compliance with their rules about who is authorized, permitted, or required to take what action. If these arrangements are not understood by public officials and public officials begin to take charge, such as has occurred over the past 25 years in the areas of forestry and irrigation, then the viability of the indigenous institution is challenged. (Decentralization: Finance and Management Project 1988, 2–3)

In funding various irrigation systems, government agencies often have their own priorities, which may not match the actual conditions of individual community irrigation systems. In many cases, the level of funding is too low to solve problems encountered by farmers in these systems. Sometimes, agency officials pay more attention to initial construction and rehabilitation than to subsequent maintenance. In other cases, the agency has its own budgetary cycle; funding for rehabilitation and maintenance is frequently released at times such as planting periods, when farmers are least able to use it effectively (see Fowler 1986).

Given the importance of community irrigation systems in many countries, it is important to examine how farmers in these systems organize themselves and what factors motivate them to cooperate in investment and water allocation. Only after policy makers have acquired this kind of knowledge can they design appropriate rules and processes of intervention.

The Plan of This Book

Although farmers can play a significant role in solving collective-action problems in both community and bureaucratic irrigation systems, potential pitfalls exist if insufficient attention is paid to factors that affect farmers' incentives for cooperation. One way to understand how different configurations of factors shape the structure of incentives facing irrigators is to examine cases in natural settings. In the past two decades, extensive case studies on both community and bureaucratic irrigation systems in various parts of the world have been written in such disciplines as anthropology, sociology, agricultural economics, and political science. These cases provide a wide diversity of experiences from which to analyze how institutional arrangements affect collective action in irrigation systems.

To learn from empirical studies of irrigation systems, one must develop a theoretical framework to identify key variables that affect collective action in these systems. One can then analyze the relationships among these variables from the perspective afforded by experiences in varied empirical settings. Knowledge of this kind can help in the design of institutional arrangements to resolve collective-action problems in specific situations.

In this book, I apply a mode of institutional analysis that identifies how different contextual variables structure various collective-action situations in irrigation systems. I examine three sets of contextual variables and their effects on human interactions: (1) the physical attributes of the resource, (2) the attributes of the community of participants, and (3) the set of institutional arrangements used (Kiser and Ostrom 1982; Oakerson 1986; E. Ostrom 1986). These contextual variables affect collective action by shaping the structure of incentives that participants face.

Based on transaction cost economics, I argue that institutional arrangements facilitate collective action by creating constraints among participants. If these constraints are effectively enforced, they help to reduce the level of uncertainty faced by the participants when they attempt to develop credible commitments and long-term cooperative arrangements with one another (North 1990; Williamson 1985). These theoretical arguments are elaborated in Chapter 2.

In Chapter 3, I discuss the research method used for this study, some problems of terminology related to irrigation, and the profiles of the case studies used in the analysis. In Chapter 4, I examine the patterns of collective outcomes in the case studies. I also analyze how these outcomes are related to various physical and community attributes. In Chapter 5, I examine some major institutional arrangements found in the case studies. I then analyze the factors that lead to the emergence or adoption of these arrangements and demonstrate how these arrangements affect the patterns of outcomes under various circumstances. In particular, I compare the performance of community and bureaucratic irrigation systems and explain how institutional arrangements and other factors can account for their differences. In Chapter 6, the concluding chapter, I discuss several theoretical propositions derived from this study. I then analyze the potential and limitations of local, self-governing organizations in the governance of irrigation systems. Finally, I examine how bureaucratic and local, self-governing organizations can work together in a mutually productive manner.

CHAPTER TWO

An Institutional Analysis of Irrigation Systems and Transaction Costs

Institutional arrangements can facilitate or impede the problem-solving capabilities of participants in irrigation systems. To learn from empirical studies of the performance of various institutional arrangements in irrigation systems, one must draw on and extend a theoretical framework that identifies the key attributes shared by collective-action situations in a wide diversity of irrigation systems. These attributes should be treated as variables that take on different values according to their specific circumstances. Relationships among these variables can then be explored in reference to experiences in varied settings. Knowledge gained in this way can assist in the diagnosis of potential problems in specific situations and the design of institutional arrangements to solve them.

Drawing on literature in political science, economics, anthropology, game theory, and law, scholars have developed a general framework of institutional analysis and development (IAD) that identifies the key working parts of typical situations facing participants in various circumstances (Kiser and Ostrom 1982; Oakerson 1986; E. Ostrom 1986). The focal point of the IAD framework is the action situation, in which individuals adopt actions or strategies. Depending on such factors as the number of participants involved, the choices available to participants, and the incentives faced by participants, different outcomes may result from interactions among participants. Many collective-action problems in irrigation systems resemble situations in which individuals trying to advance their interests end up producing unintended and harmful consequences for themselves as well as for others. One example discussed in the introductory chapter was the situation in which every farmer tried to be a "free rider" in the water allocation process and ended up worse off than if everyone had cooperated.

As in the IAD framework, transaction cost economics adopts transactions, which resemble action situations, as the fundamental unit of analysis (Williamson 1975; 1985).[1] Both transaction cost economics and institutional analysis are concerned with identifying appropriate institutional arrangements that can counteract perverse incentives inherent in various transaction situations. Whereas transaction cost economics approaches the problem by examining the characteristics of different transaction situations, the IAD framework explicitly identifies a higher level of analysis by delineating the contextual attributes that shape various action situations. At the contextual level of analysis, one examines how rules, physical attributes, and attributes of community shape various action situations.

In this chapter, I first highlight the basic theoretical premises of transaction cost economics. I then discuss how the IAD framework helps to develop more specific arguments about conditions that induce irrigators to develop and sustain various types of institutional arrangements that enable them to operate and maintain an irrigation system.

Transaction Costs and Institutional Arrangements

There are situations in irrigation when it is in everyone's interest to do one thing but they frequently end up doing something else. In such situations, each person pursuing his or her own short-term interests ends up producing suboptimal outcomes. Empirically, one finds that farmers in some irrigation systems are able to change the structures of these situations to ameliorate the perverse incentives they face while others are not. To account for this variance in outcome, an analytical approach is needed to explain both the presence and absence of cooperative behavior. To do this, one must posit a consistent model of the individual that can be used to generate predictions about likely behavior given the structure of incentives and opportunities that individual faces.

An individual's choice of action in any particular situation depends on how he or she weighs the benefits and costs of various alternatives and their likely outcomes. In an attempt to pursue benefits, an individual is, however, constrained by a limited information-processing ability. In other words, individual rationality is bounded. In many economic models, for the sake of simplification, an individual is assumed to be able to process all the information relevant to a decision situation. The individual is assumed to be able to undertake all necessary computations to reach a decision that could maximize his or her expected utilities. This assumption has been challenged by many. Herbert Simon, for example, argues that human behavior is "intendedly rational but only limitedly so" (1961, xxiv). Because the information-processing capabilities of humans are limited, individuals frequently make decisions without considering all the possible alternatives and their likely outcomes. Organizations, Simon argues, compensate for this human limitation by assigning each individual a limited task environment and standard operating procedures. Institutions that regulate ways of undertaking activities can also be considered stores of acquired knowledge. In the words of Richard Langlois:

> Institutions have an information-support function. They are, in effect, interpersonal stores of coordinative knowledge; as such, they serve to restrict at once the dimensions of the agent's problem-situation and the extent of the cognitive demands placed upon the agent. (1986, 237)

To develop mutually beneficial arrangements in irrigation, participants need rudimentary information about the physical and technological characteristics of the water flow and water delivery facilities as well as information about the respective preferences of individual participants. The information they possess at any given time and their ability to gain more information affect their ability to develop appropriate institutional arrangements to tackle their problems in water allocation and maintenance.

The long-term viability and performance of a set of institutional arrangements also depend on its ability to process information necessary for effective operation and maintenance. F. A. Hayek (1948) argues that a major economic problem of society is the continuous need to use information about the circumstances of specific time and place. In irrigation, effective water allocation and maintenance require knowledge about the topography, soil types, and crop patterns of the particular area. It is important to ensure that these kinds of information are utilized in decisions regarding water allocation and maintenance.

Opportunism, defined as "self-interest seeking with guile," is another important individual attribute that affects collective action in irrigation systems (Williamson 1985, 47). Opportunism, in conjunction with bounded rationality, creates difficulties in both negotiating and enforcing cooperative agreements. Individuals may hinder the process of negotiation by trying to hide their true preferences from one another in order to secure a better deal. After they have entered into some form of mutually agreed contract, disputes may arise as to its proper interpretation when novel situations appear or new individuals become involved. Such disputes are especially likely to occur because it is impossible to devise rules that take into account all possible future contingencies. Furthermore, individuals who have entered into a contract may still be inclined to take advantage of their fellow contractors if circumstances allow them to do so.

Transaction cost economics focuses on the disputes that may arise when individuals, who are characterized by bounded rationality and opportunism, enter into contractual relationships (Williamson 1975; 1985). Williamson argues that, contrary to the assumption of "legal centralism," which holds that the resolution of these disputes requires adjudication by an external authority, most disputes can be avoided by recognizing "potential conflict in advance and [devising] governance structures that forestall or attenuate it" (Williamson 1985, 29). These governance structures represent institutional arrangements that participants voluntarily adopt in order to foster credible commitments and to facilitate recurrent transactions among themselves. The organizational imperative that emerges from a consideration of bounded rationality and opportunism, according to Williamson, is: "Organize transactions so as to economize on bounded rationality while simultaneously safeguarding them against the hazards of opportunism" (1985, 32).

The literature of transaction cost economics studies contractual problems mostly in relation to the exchange of private goods (that is, goods characterized by the ease of exclusion and the subtractibility of resource units) such as labor and machinery (see Joskow 1988; Putterman 1986). Williamson (1985) distinguishes three principal dimensions of a transaction that are related to different organizational problems. First, some transactions are characterized by asset specificity, which "refers to durable investments that are undertaken in support of particular transactions, the opportunity cost of which investments is much lower in best alternative uses or by alternative users should the original transaction be prematurely terminated" (Williamson 1985, 55). Transactions that involve durable and transaction-specific assets experience "lock-in" effects that make unified ownership (vertical integration) commonly preferable to autonomous trading on the open market.

Second, some transactions are subject to uncertainty caused by their environments and their participants' opportunistic behavior. This uncertainty, if accompanied by significant numbers of transaction-specific assets, induces participants to devise institutional arrangements capable of sequential adaptation. Third, some transactions are undertaken more frequently than others. If a certain kind of transaction is needed only infrequently, it may not be cost-effective to establish elaborate institutional arrangements to handle it, even if transaction-specific assets are involved. On the other hand, specialized institutional arrangements will be more cost-effective if large, recurring transactions are involved.

Among the three dimensions Williamson suggests, asset specificity is the most important for transaction cost economics. He writes:

> Asset specificity is the big locomotive to which transaction cost economics owes much of its predictive content. Absent this condition, the world of contract is vastly simplified; enter asset specificity, and nonstandard contracting practices quickly appear. (1985, 56)

In irrigation, highly transaction-specific assets are involved: Once constructed, irrigation facilities such as dams or canals can hardly be relocated or redeployed for other uses. In some arid areas, farmland is a highly transaction-specific asset whose value depends on the effective functioning of an irrigation system. If an irrigation system is used by many individuals, one individual's opportunistic behavior can affect others considerably. Because it is difficult to redeploy one's investment in an irrigation system, specific institutional arrangements are essential to ensure that no one can "free ride" on others in investment and water allocation.

Uncertainty is another important factor affecting collective action in irrigation systems. One cause of uncertainty may be lack of trust among irrigators. If irrigators cannot trust one another, it is difficult for them to develop and sustain cooperative arrangements. Uncertainty may also result from external factors that are beyond the immediate control of the participants. In watercourses within some large government irrigation systems, for example, the amount of water available may vary unpredictably, subject to arbitrary decisions made by government officials operating at the system level. Uncertainty often creates obstacles for exchange and cooperation among farmers. Institutional arrangements that reduce uncertainty among participants facilitate collective action.

The frequency of transactions also affects the choice of institutional arrangements. For example, irrigation systems that require extensive cooperative efforts in maintenance may require elaborate input rules to specify and coordinate contributions from individual irrigators. Systems that require only occasional maintenance may need no specific input rules.

To understand better how these three transactional dimensions affect governance in irrigation systems, one must systematically analyze the contextual attributes that shape various transactional (action) situations. According to the IAD framework mentioned earlier, three sets of contextual attributes structure the action situation facing participants in an irrigation system: (1) the physical attributes of the irrigation system, (2) the attributes of the community of participants, and (3) the set of institutional arrangements in use by participants (see Kiser and Ostrom 1982). These three sets of attributes combine to create different incentives and constraints for participants in different systems. Because participants are characterized by bounded rationality and opportunism, they react according to the incentives and constraints inherent in the situations they face. The strategic interactions among participants in different action situations therefore produce different outcomes (see Figure 2.1).

Outcomes for participants in irrigation systems include: (1) the water supply in the system does or does not meet the water requirements of the crops in the established fields served by the system,

FIGURE 2.1   A Framework for Institutional Analysis

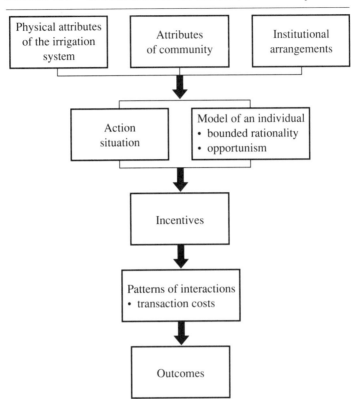

SOURCE: This and all subsequent figures and tables are devised by the author.

(2) most participants do or do not follow rules in use in the system, (3) the water diversion and delivery facilities are or are not well maintained, and (4) some participants have or have not been consistently disadvantaged in relation to the system.[2]

Some of these outcomes are influenced by the extent to which participants cooperate in the operation and maintenance of a system. The active cooperation of cultivators is essential for maintaining water diversion and delivery facilities. Whether these facilities are well maintained reflects the degree of cooperation among cultivators. Operational rules are important means of coordinating water allocation and maintenance. The extent to which most cultivators are willing regularly to follow them reflects the viability of these rules as coordinating devices.

In some situations, however, outcomes may be beyond the immediate control of the participants. The volume of water flow in a river at any particular moment, for example, is frequently a result of uncontrollable physical and meteorological factors.

Different outcomes may be related to the different action situations faced by participants. The outcomes of one action situation may form the context of another situation. The level of water supply in a system may be partly a result of the way participants constructed the system in the first place. Once the system has been constructed, the amount of water supplied by it, as will be discussed in Chapter 4, becomes a physical attribute that affects irrigators' incentives to cooperate in maintaining the system.

Individuals in different situations may evaluate these outcomes differently. For example, irrigators in general are likely to regard adequate water supplies and good maintenance as the most important outcomes. Different communities, however, may form different evaluations of the fourth outcome (some participants have or have not been consistently disadvantaged), depending on the concept of fairness their members share. Government and international development agencies frequently rate this outcome relatively high when designing their irrigation projects. Evaluation of the second outcome (most participants do or do not follow the rules in use) may also differ among participants, depending on whether they regard the rules in use as legitimate.

In the remaining part of this chapter, I first discuss how physical and community attributes affect collective action and outcomes in an irrigation system. I then examine how institutional arrangements affect the structures of incentives faced by participants and how participants relate to one another.

Physical and Community Attributes

As discussed in Chapter 1, most irrigation systems are characterized by the difficulties of exclusion and the subtractibility of resource units. These physical attributes create collective-action situations among irrigators in most systems. Additional physical attributes, including the size of the irrigation system, the pattern of water supply, and the availability of alternative water sources, also affect interactions among irrigators. Community attributes such as irrigators' sources of incomes and the presence or absence of social, economic, cultural, and locational differences among irrigators also affect the irrigators' incentives to cooperate.[3] Either by itself or in combination with others, each of these attributes potentially affects collective action and outcomes in an irrigation system. The constraints and opportunities thus created must be taken into account in the design of institutional arrangements for an irrigation system.

Dependence on the System. The degree to which farmers depend on an irrigation system may affect their incentives to cooperate in fairly complex and counterintuitive ways. Farmers may depend on an irrigation system in two different senses: (1) as a major source of income (that is, their incomes derive mostly from cultivating crops irrigated by the system); and (2) as a major source of water for irrigation.

The extent of farmers' dependence on an irrigation system as a major source of income may have different effects on their incentives to participate in collective action. In most situations, the more irrigators depend on an irrigation system, the more likely they are to expend substantial private resources to operate and maintain the system. Irrigators without other job obligations are also more likely to be able to participate in collective activities in an irrigation system.

In other situations, if most farmers do not have other sources of income, it may be difficult for them to develop new cooperative ventures that require substantial capital investment or sacrifice before producing benefits. It may, for example, require a substantial reduction in the rate of water withdrawal to replenish a deteriorating water basin. If most farmers rely entirely on water from the basin to irrigate their crops and have no alternative source of income, it will be difficult for them to develop and enforce cooperative efforts to cut back on the rate of water withdrawal. Depending on the circumstances, farmers' dependence on an irrigation system as a major source of income may either facilitate or impede collective action.

By the same token, the availability of alternative water sources may increase or decrease farmers' incentives to cooperate. In some situations, the availability of an alternative source of water may reduce tension among irrigators when water flow in the system is scarce, thus facilitating their long-term cooperation. In other situations, irrigators with access to an alternative source of water may be less willing to contribute to the operation and maintenance of the system than those without, thus inhibiting their long-term cooperation.

Water Scarcity and Uncertainty.   Farmers' vulnerability to scarcity and uncertainty in water supply and its effects on their incentives for collective action have drawn special attention in the irrigation literature. Wickham and Valera (1979), in a study of irrigation projects in the Philippines, observe that in order to induce farmers to cooperate in managing their watercourses, an effective systemwide management program is a prerequisite. In other words, farmers have less incentive to organize if they do not have a predictable or sufficient flow of water into their watercourses in the first place. This observation seems to contradict that of Wade (1988a) who, drawing upon experiences in south India, argues that the greater the scarcity and uncertainty of the water supply, the greater the likelihood that a community of cultivators will develop collective arrangements to govern their watercourse.

Although these two arguments appear to be directly contradictory, they may be consistent when presented in a more general context. Irrigators' vulnerability to scarcity and uncertainty in water supply may be related in a curvilinear manner to their incentives to cooperate (see Uphoff, Wickramasinghe, and Wijayaratna 1990). Farmers have to be sure of at least some minimal availability of water before they are willing to invest in collective efforts in water allocation and maintenance. If the water supply is abundant, however, investments in water allocation and maintenance would make little sense as water will be available anyhow. But under conditions of moderate scarcity, keeping regular water allocation and maintenance schedules may strongly affect the amount of water available to farmers' fields. Thus little collective action by farmers can be expected under conditions of either extreme abundance or scarcity. Most collective activities occur in situations where water is barely sufficient or moderately scarce and farmers believe that their collective efforts can improve their chance of securing a more reliable supply.[4]

An inadequate supply of water, however, could increase coordination costs among farmers. As the supply decreases, the temptation for free riding in water acquisition increases; efforts in monitoring and sanctioning have to be increased to enforce discipline in water allocation. In addition, more conflicts are likely to arise among irrigators as they compete for a scarce source of water. In some situations, farmers may be able to increase the water flow to their fields by, for instance, damaging the canal embankment. This again increases the difficulty of maintaining the irrigation system. All these incidents could increase the costs of organizing collective action in irrigation.

Thus, in situations between extreme abundance and extreme scarcity, farmers expect both potential benefits and costs in their participation in collective action. On the one hand, if they are successful in collective action, they may be able to receive a more adequate and reliable supply of water; there is a "demand" for collective action. On the other hand, the potential costs created by water scarcity make their cooperation with one another more difficult, thus inhibiting the "supply" of collective action. One may expect that in the real world many irrigation systems fall within this middle range; whether or not farmers in these systems are successful in governing and maintaining their systems depends on the balance between the benefits and costs they face.

Irrigated Area and Number of Irrigators. Even if individual irrigators are willing to contribute to collective endeavors, they have to expend resources to organize among themselves to assign responsibilities and undertake water allocation and maintenance jobs. Both the size of the irrigation system and the number of users of the system may affect farmers' actions. Many authors argue that, all other things being equal, information gathering, communication, decision making, and monitoring costs increase as the size of a resource increases. By the same token, various kinds of transaction costs increase as the number of irrigators increases (Field 1986; Buchanan and Tullock 1962). These two arguments imply that, all other things being equal, it will be easier to organize collective action in irrigation systems of smaller sizes and with fewer users.

Although it is more costly to organize collective action in large irrigation systems, this does not mean that large systems are doomed to fail. In many circumstances, in order to take advantage of a large source of water, it is more economical to develop a system that irrigates extensive areas and serves many farmers. Depending on the kinds of institutional arrangements adopted, coordination problems in large systems can be solved in various manners.

The type of institutional arrangement that is needed to overcome the problem of organizing large-scale irrigation has long been of interest to social scientists. Wittfogel's thesis that large-scale irrigation (hydraulic agriculture) requires discipline and direction by an external authority is probably the most famous theory about irrigation known to a general social science audience.[5] Wittfogel writes:

> A large quantity of water can be channeled and kept within bounds only by the use of mass labor; and this mass labor must be coordinated, disciplined, and led. Thus a number of farmers eager to conquer arid lowlands and plains are forced to invoke the organizational devices which—on the basis of premachine technology—offer the one chance of success: they must work in cooperation with their fellows and subordinate themselves to a directing authority. (1981, 18)

Wittfogel further argues that the need to direct and enforce cooperation in constructing and operating major hydraulic works induced the development of highly centralized bureaucratic regimes in many parts of the world.

This thesis, however, has been contradicted by many examples in which farmers or local communities have been able to assemble and discipline massive local labor and other resources to construct and sustain irrigation systems with command areas of over several hundred hectares (for example, Lando 1979; Siy 1982; Pradhan 1983). These systems are not governed by any single, unified bureaucratic machinery. Instead, federated arrangements are adopted under which the entire system is governed by multiple layers of farmers' organizations. As will be discussed later in this chapter, this kind of multilevel arrangement can reduce transaction costs and facilitate coordination and problem solving in large irrigation systems.

Differences among Irrigators. Irrigators differ from one another in (1) their cultural and social characteristics such as ethnicity, caste, race, clan, or religion; (2) the amounts of irrigated land or water shares they hold; or (3) the locations of their plots within the system. These differences are important contextual attributes that affect collective action in irrigation.

If a community of irrigators is divided by ethnic, clan, racial, caste, or religious differences that inhibit communication, the costs of organizing collective action within the community will be higher than within communities without divisions. In some situations, the divisions among irrigators may be great enough to inhibit any form of cooperation. Cases exist, however, in which communities with ethnic, caste, or other divisions are able to overcome these obstacles and develop and sustain long-term cooperative arrangements. In these situations, high levels of potential disagreements and conflicts among irrigators still exist. Institutional arrangements that can mitigate and resolve potential conflicts among farmers and ensure a more equitable sharing of benefits and burdens help to sustain their cooperative efforts.

Some literature suggests that a collective good is likely to be provided if a few individuals have disproportionate interests in the good, since these individuals have more to gain from the good and may find it in their own interests to provide the good by themselves or expend resources to organize other potential beneficiaries to provide the good (for instance, Olson 1965). In irrigation, this means that the presence of individuals with disproportionate landholding or shares of the water flow facilitates collective efforts in water allocation and investment. Conversely, some authors argue that a highly unequal distribution of landholding inhibits local cooperation in operating and maintaining irrigation facilities (for example, Palanisami and Easter 1986). Farmers with disproportionate wealth and influence may be reluctant to cooperate with poorer farmers; or if they do, they expect more privileges and benefits (Harriss 1977). To ensure a mutually productive relationship among all farmers in this kind of situation, it is important to have institutional arrangements that ensure a fair sharing of costs and benefits among participants.

Irrigators may have unequal access to the flow of water. This difference also affects their incentives for cooperation. In most canal irrigation systems, headenders have a natural advantage over tailenders in their access to water. As documented by many authors, unless irrigation systems are well organized, headenders tend to take more water than is necessary for the growth of their crops, to the detriment of tailenders (Bromley 1982; Chambers 1977). The temptation to "overuse" water is especially great for the cultivation of rice. Rice is believed by many farmers to be very sensitive to water shortage but tolerant to large amounts of water.[6] Standing water is also an important means to control the growth of weeds. For many farmers, to maintain as much water as possible in their rice field is a good way to reduce the risk of lower yields and the amount of labor required to clear weeds (see Abel 1977). Because of their more favorable location, headenders may have little incentive to cooperate with tailenders in water allocation.

The position of headenders is, of course, not invulnerable. Tailenders may go upstream and hurt headenders by destroying their banks, gates, or valves if no one sanctions them. The possibility that their diversion works will be destroyed induces headenders to cooperate with tailenders to a certain extent. On the other hand, when both headenders and tailenders implement a set of enforceable allocation rules, headenders are probably in a better position to negotiate a more favorable share of water because of their proximity to the source.

The situation is different in irrigation systems where most farmers cultivate plots in both head and tail areas. In this kind of system, most farmers have vested interests in ensuring that enough water is delivered to the tail area; this pattern of plot distribution can facilitate cooperation. In some irrigation systems, specific rules exist to make sure that every farmer cultivates plots in both the head and tail areas (Coward 1979).

Conclusion. Physical and community attributes create the setting in which irrigators interact. Although many of the physical and communal attributes of an irrigation system affect the situations that irrigators face, few of these attributes have deterministic effects on the success or failure of collective action. In some cases, institutional arrangements can mitigate the perverse effects of situations created by these attributes.

Institutional Arrangements

From a policy perspective, institutional arrangements are the most important of the three contextual attributes underlying action situations faced by irrigators. Institutional arrangements are rules that "are potentially linguistic entities that refer to prescriptions commonly known and used by a set of participants to order repetitive, interdependent relationships" (E. Ostrom 1986, 22). In a rule-structured situation, individuals select specific actions from a large set of allowable actions in light of the incentives existing in the situation. Rules as social artifacts are subject to human design and intervention. By identifying the capabilities and limitations inherent in different institutional arrangements, one can anticipate different patterns of social outcomes. By changing rules, one can intervene to change the structure of incentives for participants and the relationship among participants. Such interventions may enhance or reduce irrigators' abilities to allocate water and maintain an irrigation system effectively.

Operational Rules. Operational rules define who can participate in which situations; what the participants may, must, or must not do; and how they will be rewarded or punished. Operational rules facilitate coordination if the participants share a common knowledge of these rules and are willing to follow them. In a world of rapidly expanding knowledge and changing circumstances, rules have to be able to create enough predictability among individuals yet permit enough flexibility to deal with various contingencies (V. Ostrom 1989). In irrigation systems, four kinds of operational rules are particularly important if farmers are to solve their collective-action problems.[7] These include boundary rules, allocation rules, input rules, and penalty rules.

Boundary rules. A key precondition for successful collective action in common-pool resources is the effective enforcement of a set of boundary rules that limits the number of individuals entitled to resource units (E. Ostrom 1990; Schlager and Ostrom 1992). Without a well-defined set of rights holders, it is difficult for actual and potential users to negotiate and enforce a common set of rules coordinating various water allocation and investment activities. Arthur Maass and Raymond Anderson, for example, argue that "the strength and coherence of local irrigation organizations in developed regions appears to be correlated with an irrigation community's success in limiting or stabilizing growth, thereby gaining security for its members" (Maass and Anderson 1986, 368).[8]

The existence of a closed set of rights holders also distinguishes a common-property resource from an open-access resource (Bromley 1984). Norman Uphoff suggests that because the resource and the users are more clearly defined, water user associations tend to outperform other local organizations responsible for such resources as forests and grazing land (Uphoff 1986b, 27–28).

Several boundary requirements are frequently used in irrigation systems: (1) ownership or leasing of land within a specified location, (2) ownership or leasing of shares in water delivery facilities, (3) ownership or leasing of shares to a certain proportion of the water flow, (4) payment of certain entry fees, and (5) membership in an organization. A boundary rule may consist of only one requirement or a combination of requirements.

Although limiting the number of users of an irrigation system is a way to ensure its long-term viability, serious resource misallocation occurs if individuals who could benefit from an irrigation system are excluded. This may happen, for example, when the irrigation system has an abundant supply of water but the rights to take water are rigidly tied to plots within a certain area. Excess water will be wasted if farmers who cultivate plots outside the area are excluded from the system.

There are, however, boundary requirements that tend to encourage efficient uses of water. Some authors, for example, argue that transferable water rights, independent of land, create incentives for individuals to use water efficiently. Transferable rights also allow the trading of water shares so that water can be obtained by the individuals who can use it most productively (Martin and Yoder 1986; Anderson 1983). Others, however, argue that independently transferable rights generally require more technological and organizational control and may not be feasible in all kinds of situations (Glick 1970).

Allocation rules. Allocation rules prescribe the procedure for withdrawing water from an irrigation system. They are especially important when the supply of water is inadequate to meet the crop requirements of all cultivators simultaneously. If allocation rules are effectively enforced, they can reduce uncertainty and conflict among irrigators in relation to water withdrawal. Three types of procedures—fixed percentage, fixed time slot, and fixed order—are frequently used in water allocation:

1. *Fixed percentage:* The flow of water is divided into fixed proportions by some physical device.

2. *Fixed time slot:* Each individual is assigned fixed time slots during which withdrawal is permitted.

3. *Fixed order:* Individuals take turns getting water.

Each of these procedures may be based on different premises such as amount of land held, amount of water needed for cultivation, number of shares held, historical pattern of use, location of fields, or official discretion. An allocation rule may, for example, require each irrigator to withdraw water in specific time slots. The length of the slot for an irrigator may be determined by the amount of land he holds; that is, the more land he holds, the more time to which he is entitled. There are many possible combinations of water distribution procedures and bases.[9]

Depending on such diverse attributes as the degree of water scarcity, the length and structure of the water-carrying facilities, the types of crops cultivated, and the monitoring devices available, different allocation rules may be appropriate under various situations. Among them, the degree of water scarcity deserves specific discussion. The degree of water scarcity affects the type of allocation rules needed to coordinate water appropriation activities. In systems that have an abundant supply of water all year round, no specific allocation rule may even be needed. For many other irrigation systems, the volume of water supply as a whole may be adequate for the requirement of all the crops cultivated by its members, yet demands for water may exceed the amount available during certain time periods of the year. This situation happens frequently in dry seasons or in specific growth stages of crops when larger amounts of water are needed.

Two different responses to such a situation are possible. One possible response is to impose a more restrictive set of allocation rules: More restrictive turns or time schedules may be adopted, or officials may begin to exercise discretion in allocating water among farmers. To enforce this more restrictive set of rules, irrigation institutions and officials must be able to command sufficient respect and confidence from irrigators. Otherwise, pressure from irrigators, especially the more influential ones, may undermine the governing ability of the institutions and officials. Another possible response to decreases in water supplies is to suspend or relax restrictions on water allocation. This response may lessen pressure on the institutions and officials. Conflict may, however, develop among irrigators unless they have other sources of water on which to fall back. Furthermore, tailenders are likely to suffer more than headenders in the absence of allocation arrangements.

Input rules.   Input rules prescribe the types and amounts of resources required of each cultivator. Irrigators who own and run an entire system must raise their own resources to finance their own organization and to construct and maintain water delivery facilities. In large-scale, government-built irrigation systems, human and material resources from irrigators could also be effective and reliable inputs for developing and maintaining systems. An irrigator may be required to contribute four major types of inputs: (1) regular water tax, (2) labor for regular maintenance, (3) labor for emergency repair, and (4) labor, money, or materials for major capital investment. Each of these input requirements may be based on one of two kinds of premises—equal or proportional. Equal rules simply require equal contributions from all irrigators. Proportional rules require contributions from irrigators roughly in proportion to the benefits each gets from the system—for example, proportional to the irrigator's share of the system, to the amount of land cultivated, or to the amount of water needed.

Some scholars argue that if farmers are required to contribute labor to maintenance, the inputs required should be proportional to the benefits received. Chambers, for example, writes:

> Communal labor is most likely to be effective where the community will benefit directly and where labor obligations are proportional to expected benefits . . . . Conversely, where there is no direct link between the work done and the benefits gained, communal maintenance will be much more difficult. (1977, 354)

According to this principle, proportional input rules should be more effective for maintenance than equal input rules. There are, however, exceptions to this principle. First, if an irrigation system requires relatively small amounts of labor inputs for regular maintenance every year, the costs of implementing proportional rules could exceed their potential benefits. Only in systems that require large labor inputs could the gains from proportional rules be higher than the costs of enforcing them. Second, if an important structure, such as the diversion dam, is destroyed and requires emergency repair, equal contribution rules may be easier to implement than proportional rules. The prospect of losing the entire source of water may create enough incentives for everyone to participate in repairing the structure.

Penalty rules.   In most cases, rules will be ineffective unless rule breakers are subject to punishment. Some possible penalties include community shunning, fines, temporary or permanent loss of rights to water, and incarceration. Which of these penalities is a more effective deterrent depends on the features of the community of irrigators and the monitoring mechanisms available. In a closed and homogeneous community, shunning may be sufficient. In a more diverse and heterogeneous community, more substantial penalties such as fines are necessary. Such serious penalties as loss of rights to water and incarceration may not be suitable for every irrigation community because they may induce a high level of conflict among irrigators. Unless backed by external authority with legal power for coercion, these penalties may be difficult to enforce.

Collective-Choice Arrangements. Operational rules establish constraints that, if properly designed and followed, facilitate cooperation among participants in various collective-action situations in irrigation. Operational rules, however, are neither self-generating nor self-enforcing. In most cases, institutional arrangements must be established to adjudicate conflicts, enforce decisions, and formulate and modify operational rules. These institutional arrangements represent a second set of rules—collective-choice rules. The study of processes used to create, enforce, and modify collective-choice rules involves a different level of institutional analysis, the constitutional level (V. Ostrom 1987).

Collective-choice arrangements for determining, enforcing, and altering operational rules are especially important in view of participants' bounded rationality and opportunism. Bounded rationality makes it impossible to devise operational rules that anticipate all kinds of contingencies; disputes among participants as to the proper meaning and scope of operational rules can arise frequently.

Collective-choice arrangements structure the processes by which disputes among participants can be settled. Opportunism makes individuals inclined to take advantage of their fellow contractors; collective-choice arrangements that sanction against rule-breaking behavior are important for sustaining mutually productive relationships. Furthermore, in a world of changing knowledge and environments, operational rules adopted at one time may become obsolete at another; institutional arrangements that facilitate the adoption and modification of rules enable participants to respond to these changes.

Multiple levels of collective-choice entities. Different sets of collective-choice rules and different communities of participants may be involved in collective-choice situations. Depending on attributes such as the size and number of users of the irrigation system, different collective-choice entities may be constituted to exercise collective-choice prerogatives on behalf of the users and other concerned parties. Some irrigation systems, for example, are governed solely by a national government agency; operational rules may be created, changed, and enforced by reference to statutes adopted by the national legislature or executive. The collective-choice entity in this case involves not just one specific community of irrigators but also potential irrigators, interest groups, politicians, government officials, and the general public, who share an interest in irrigation and other related activities. In some other irrigation systems, the collective-choice entity is constituted solely by irrigators who adopt and enforce their own collective-choice and operational rules.

Sometimes a community of irrigators may be subject to multiple sets of operational rules adopted by two different collective-choice entities. For example, irrigators in large irrigation systems may be simultaneously subject to two sets of operational rules adopted by two different collective-choice entities—one entity at the system level and another at a subsystem level.[10] Collective-choice entities at the subsystem level, constituted by farmers themselves, are important for the effective operation and maintenance of large irrigation systems for two major reasons. First, what kinds of water allocation and input rules are the most effective and how these rules should be implemented depend greatly on such specific attributes as soil type, field topography, cropping pattern, and amount of water available in the specific irrigated area. Frequent, quick, but nonroutine decisions must be made about water allocation and maintenance in response to changes in such variables as volume of water flow, climate, and growth stage of plants. In many large irrigation systems, different watercourses vary in these attributes. If there is only one collective-choice entity to create and enforce one uniform set of operational rules for an entire system, the set of rules is unlikely to serve the needs of all watercourses equally well. Local collective-choice entities at the watercourse level, if properly constituted, are likely to facilitate the utilization of "information of specific time and place" in formulating and enforcing appropriate operational rules and choices (Hayek 1948).

Second, collective-choice entities at the subsystem level involve irrigators in formulating their own rules. Irrigators are more likely to have incentives to follow and enforce rules they adopted themselves than those handed down from an outside authority. Irrigators can also mobilize various informal mechanisms such as social shunning to enforce their own rules, mechanisms unavailable to external officials.

Whereas collective-choice entities at the subsystem level facilitate adaptation to the specific needs of various irrigation units, a collective-choice entity at the system level is necessary to deal with broader collective problems such as the allocation of water among watercourses and the maintenance of diversion works for the entire system. The collective-choice entities at the subsystem level, however, can still maintain their autonomy in relation to water allocation and maintenance within their respective areas. If different levels of collective-choice entities are constituted to deal with collective-action problems of different scopes, many coordination and control problems associated with large irrigation bureaucracies can be avoided.

Collective-choice rules. Individuals may have little incentive to follow rules unless they believe that their noncompliance will result in substantial punishment. Long-term cooperation among a large group of individuals depends on arrangements that help monitor and sanction against noncompliance (see Hechter 1987). Mutual monitoring among irrigators can be a means of rule enforcement. It may be effective when (1) only a small group of individuals is involved, (2) each individual's activities can be easily observed by others, and (3) each individual has an incentive to monitor others' activities in order to protect his or her own rights. When many individuals are involved, however, the provision of monitoring is itself subject to the free-riding problem because certain individuals may have incentives to save the time and energy for monitoring others' activities, hoping that others will do the monitoring job for them. Specialized officials may be needed to enforce rules.

Many cooperative activities in irrigation benefit from the involvement of specialized officials. Officials vested with special prerogatives in rule formulation and enforcement, however, are frequently in a position to interpret rules to their own advantage or demand favor from irrigators when adjudicating their disputes or distributing their water shares. This potential opportunistic behavior of officials is a permanent danger in any collective-choice entity. The design of institutional arrangements that can ensure the accountability of irrigation officials has been a major concern of the literature in irrigation organization and management (Hunt 1989; Coward 1980a; Chambers 1988; Wade 1988b).

To ensure the responsiveness of irrigation officials to irrigators, rules are needed to stipulate how irrigation officials are to be selected and removed, to whom they must report, and how they are to be compensated for their services. These collective-choice rules affect the structures of incentives faced by these officials and their services to irrigators. These officials are more likely to be responsive to irrigators' needs if (1) their tenures are subject to periodic votes by irrigators, (2) they must periodically report to irrigators in general meetings or hearings, and (3) their salaries depend on direct contributions from irrigators.

In some irrigation systems, incentives for officials come more from their private interests in the operation and maintenance of the system than from their official salaries. If the officials themselves, for example, cultivate lands in the tail end of a system, it would be in their personal interest to ensure that the water allocation and maintenance schedules are being followed by all irrigators so that their fields can get a sufficient and predictable supply of water. In this situation, personal interests are sufficient incentives for the officials to work for the common interests of the collective entity.

Conclusion. Institutional arrangements create constraints that facilitate cooperation among irrigators. No single form of institutional arrangement, however, is good for all circumstances. Different operational and collective-choice rules, in combination with the physical and community attributes of an irrigation system, may create different incentive structures that induce cooperation or conflict among participants.

Research Agenda

Various physical, community, and institutional attributes may affect collective-action situations in irrigation. These attributes usually combine in a configurational manner rather than a simple, additive manner (see Kiser and Ostrom 1982; E. Ostrom 1986). To know the effect of one attribute, one must know what other attributes are also in effect. A change in one attribute may alter the way the entire configuration operates, thus creating quite a different action situation. This principle implies that when one tries to explain or predict outcomes for various irrigation situations, one must be aware of the interrelationships among the contextual attributes involved.

This chapter's discussion of collective-action problems in terms of transaction costs helps to specify several behavioral assumptions and contextual attributes that potentially affect outcomes in an irrigation system. Figure 2.2 shows how these assumptions, attributes, and outcomes are related to one another within the framework of institutional analysis.

FIGURE 2.2   Research Agenda

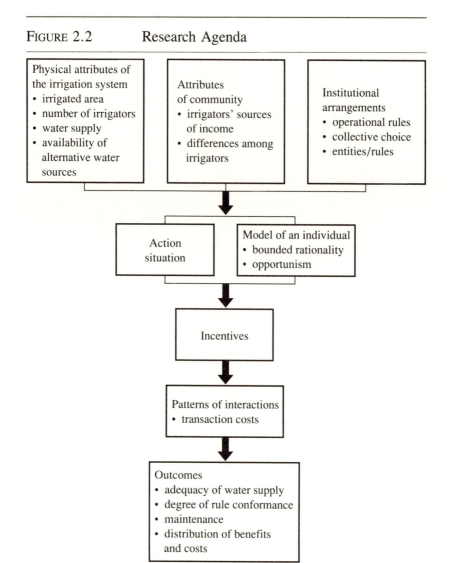

CHAPTER THREE

Comparing Irrigation Systems and Institutions

A way to ascertain how various physical, community, and institutional attributes affect the performance of an irrigation system is to examine their pattern of interactions in natural settings. In the past two decades, scholars in such disciplines as anthropology, sociology, agricultural economics, and political science have written extensive in-depth case studies on irrigation systems in various parts of the world. These cases vary from extremely simple settings, in which temporary dams divert water from streams to small, homogeneous groups of farmers to complex settings in which huge networks of canals deliver water to diverse groups of people and hundreds of thousands of hectares of farmland. Although some of these studies focus on certain selected aspects of an irrigation system, they represent in many instances excellent accounts of how different physical, community, and institutional attributes affect the process of organizing various types of collective action related to irrigation systems. Information from some of these case studies is used to examine arguments discussed in Chapter 2.

In this chapter, I first introduce some basic terminology essential for classifying and comparing irrigation systems. I then describe how case studies on irrigation systems have been collected and used for analysis in the subsequent chapters of this book.

Simple and Complex Irrigation Systems

How the boundaries of an irrigation system are defined determines how one identifies such important variables as the size of the system, the number of irrigators, and the institutional arrangements related to the system.[1] Unless definitions are used consistently, the validity of any comparative studies will be in doubt. One way of conceptualizing the boundaries of an irrigation system is to consider its water delivery processes.

These processes can be divided into four stages or distinct resource parts—production, distribution, appropriation, and use (see Plott and Meyer 1975).[2] The production of water for irrigation involves making water available at locations and times when it does not naturally occur in the form of precipitation and immediate runoff. Water is produced, for example, by damming the flow of a river and releasing it during irrigation seasons. A dam or any other form of headwork is the *production resource* of the irrigation system. From the production resource, the water may be distributed through a large aqueduct or canal to the irrigated area; the aqueduct or canal is the *distribution resource*. In the irrigated area, farmers may appropriate water from the local canals, tanks, or pumps; these structures are the *appropriation resources*. The water appropriated by farmers is then used to irrigate crops in fields; the fields and crops together constitute the *use resources*.[3]

Although appropriation resources can be distinguished from production and distribution resources in many irrigation systems, in other systems the appropriation and distribution resources may be contained in the same boundary. If, for example, water is diverted to the fields immediately after it leaves the headwork, the network of canals connected to the headwork can be considered both the distribution and the appropriation resources, which are in this case identical.

With this distinction among production, distribution, and appropriation resources, one can identify two general types of irrigation systems—simple and complex. In a simple irrigation system, the production and distribution resources supply water to only one appropriation area. In a complex irrigation system, the production and distribution resources deliver water to multiple appropriation areas (see Figure 3.1).[4]

Simple irrigation systems are generally easier to analyze because

FIGURE 3.1   Simple and Complex Irrigation Systems

River   Dam (production resource)

Main canal (distribution resource)

A network of smaller canals (appropriation resource)

Fields and crops (use resource)

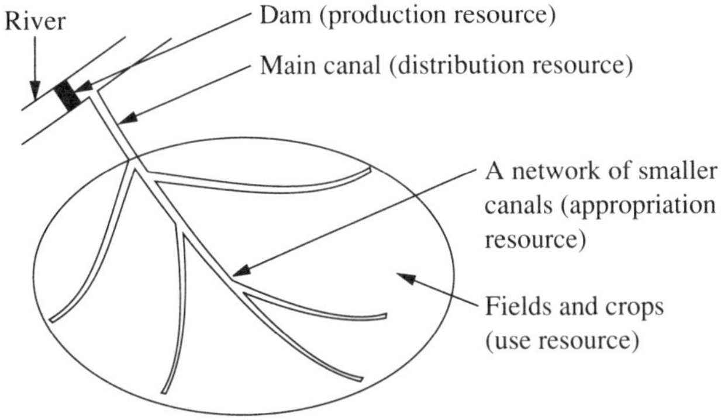

(*a*) *A Typical Simple System*

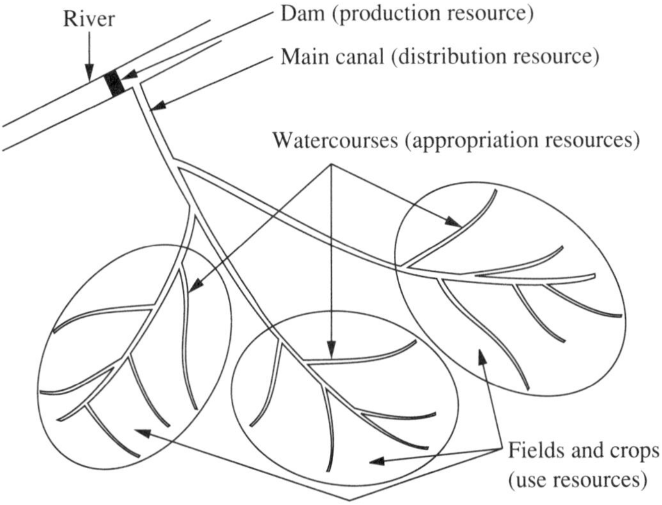

River   Dam (production resource)

Main canal (distribution resource)

Watercourses (appropriation resources)

Fields and crops (use resources)

(*b*) *A Typical Complex System*

the entire network of canals usually constitutes the appropriation resource and the organizational activities of all irrigators center on it. Analytical problems arise, however, in regard to complex irrigation systems that are divided into many smaller watercourses (appropriation areas). Although problems at the system level certainly affect various appropriation areas within the system, each appropriation area has its own set of collective-action problems.

Other problems arise when one attempts to compare simple and complex irrigation systems that may be physically, technologically, and institutionally different from each other. To solve these problems, I use individual appropriation resources as the units for comparison in this study. I focus on the appropriation resource because, regardless of the institutional arrangement of an irrigation system, irrigators are always involved in the appropriation stage of the water delivery process. It is also at this stage that many water allocation and maintenance problems arise.[5] In subsequent chapters, I analyze activities and attributes related to the entire appropriation areas of simple irrigation systems and selected appropriation areas (watercourses) within complex systems. The activities and attributes related to production and distribution resources are addressed whenever they closely relate to activities within the appropriation area under discussion.

Organizational Forms

An irrigation system may be governed by one or more collective-choice entities, which may be constituted by (1) a national or regional government agency or enterprise, (2) a local government unit, (3) a communal enterprise or an irrigators' association, or (4) any other kind of organization such as a profit-making private enterprise. In some irrigation systems, one collective-choice entity governs the production, distribution, and appropriation resources simultaneously. In some other systems, a separate collective-choice entity governs each of the three resources. An irrigation system can be classified according to the kinds of collective-choice entities involved in governance and the kinds of resources governed.

In this study, I concentrate on two kinds of irrigation systems— bureaucratic and community. In a bureaucratic irrigation system, the production resource is governed by a national or regional government agency or enterprise. In some bureaucratic systems, the same government agency or enterprise may also govern the distribution and appropriation resources of the system. In others, different collective-choice entities, such as irrigators' associations, may be involved in governing the distribution or appropriation resources.[6]

In a community irrigation system, the production resource is governed by either communal enterprises or irrigators' associations.

In almost all community irrigation systems, the distribution and appropriation resources are also governed by either communal enterprises or irrigators' associations. Systems that are not governed by any formal collective-choice entity are also considered community irrigation systems since only irrigators are involved in their operation.

The Nature of the Evidence

The data for this study were collected through a research project on the study of common-pool resources conducted by the Workshop in Political Theory and Policy Analysis at Indiana University. A part of the research project has been to undertake a systematic analysis of in-depth case studies of common-pool resources, including irrigation systems, fisheries, forests, groundwater basins, and grazing land.[7] Members of the research project have developed a series of in-depth coding forms, containing mostly closed-ended questions, to identify the key physical, community, and institutional attributes of an appropriation area in a common-pool resource and to obtain general information about production and distribution resources when they are separated geographically and organizationally from the appropriation resource. Forms that are relevant to this study include the following:

1. The *location form* examines the major geographic and demographic features of the location of an appropriation resource.

2. The *appropriation resource form* examines the boundaries and physical characteristics of an appropriation resource.

3. The *operational-level form* examines the types of situations faced by participants, the level of information available to them, their potential actions and levels of control, their patterns of interaction, and the outcomes they obtain.

4. The *subgroup form* examines the stakes and resources, potential actions and levels of control, and strategies of participants in a subgroup. There is always more than one

subgroup, and more than one subgroup form has to be filled out if the participants in an appropriation resource do not have relatively symmetrical legal rights to appropriate water, withdrawal rate from the resource, exposure to variation in water supply, level of dependency on water from the resource, and patterns of use.

5. The *operational rule form* examines the kinds of boundary, authority, scope, information, payoff, and aggregation rules used in an appropriation resource.

6. The *collective-choice form* examines the collective-choice entities that govern an irrigation system.

7. The *organizational structure form* examines the structure and process of a collective-choice entity. Multiple organizational structure forms have to be filled out if more than one collective-choice entity is involved in governing an appropriation resource.

Most of the variables discussed in Chapter 2 are covered by these seven forms. I have used these forms to code data provided by in-depth case studies.

The research project has undertaken extensive efforts to identify theoretical and case studies in irrigation systems and other common-pool resources. Over 1,000 items, including books, dissertations, journal articles, monographs, and occasional papers have been identified in the area of water resources and irrigation (F. Martin 1989), and over 450 documents have been collected by the research project. Cases were selected from these documents for coding only if they contain detailed information about (1) participants in the resource, (2) strategies used by participants, (3) the condition of the resource, and (4) rules in use for the resource. Cases were also selected so as to include in the sample as much diversity in physical, community, and institutional attributes and collective outcomes as possible.

This study is based on forty-seven coded cases. The profiles of these cases are shown in Tables 3.1, 3.2, and 3.3. Twenty-nine of these cases are community systems; fourteen are bureaucratic systems; and four are systems governed by local governments.

TABLE 3.1　　Community Irrigation Systems: Cases Coded

Country	System name	System type	Command area (in hectares)	Major crop	Documentation
Bangladesh	Nabagram	Simple	29	—	Coward & Badaruddin (1979)
Indonesia	Bondar Parhudagar	Simple	4	Rice	Lando (1979)
Indonesia	Takkapala	Simple	95	Rice	Hafid & Hayami (1979)
Indonesia	Saebah	Simple	100	Rice	Hafid & Hayami (1979)
Indonesia	Silean Banua	Simple	120	Rice	Lando (1979)
Iran	Deh Salm	Simple	300	Other grains	Spooner (1971, 1972 & 1974)
Iran	Nayband	Simple	—	Rice	Spooner (1971, 1972 & 1974)
Nepal	Raj Kulo	Simple	94	Rice	Martin & Yoder (1983a, 1983b & 1986)
Nepal	Thulo Kulo	Simple	39	Rice	Martin & Yoder (1983a, 1983b & 1986)
Nepal	Char Hazar	Simple	200	Rice	Fowler (1986)
Nepal	Chhahare Khola	Simple	20	Other grains	Water & Engineering Commission (1987)
Nepal	Naya Dhara	Simple	55	Rice	Water & Engineering Commission (1987)
Philippines	Agcuyo	Simple	9	Rice	de los Reyes et al. (1980a)
Philippines	Cadchog	Simple	3	Rice	de los Reyes et al. (1980a)
Philippines	Calaoaan	Simple	150	Rice	de los Reyes et al. (1980a)
Philippines	Mauraro	Simple	15	Rice	de los Reyes et al. (1980a)
Philippines	Oaig-Daya	Simple	100	Rice	de los Reyes et al. (1980a)
Philippines	Sabangan Bato	Simple	94	Rice	de los Reyes et al. (1980a)
Philippines	Silag-Butir	Simple	114	Rice	de los Reyes et al. (1980a)
Philippines	San Antonio 1	Simple	23	Rice	de los Reyes et al. (1980b)
Philippines	San Antonio 2	Simple	7	Rice	de los Reyes et al. (1980b)
Philippines	Tanowong T	Simple	—	Rice	Bacdayan (1980)
Philippines	Tanowong B	Simple	—	Rice	Bacdayan (1980)
Philippines	Pinagbayanan	Simple	20	Rice	Cruz (1975)
Tanzania	Kheri	Simple	260	Other grains	Gray (1963)
Thailand	Na Pae	Simple	64	Rice	Tan-kim-yong (1983)
Philippines	Zanjera Danum Sitio	Complex	45/1500[a]	Rice	Coward (1979)
Switzerland	Felderin	Complex	19/—	Meadow	Netting (1974 & 1981)
Thailand	Chiangmai	Complex	—/—	Rice	Potter (1976)

N = 29
— = Missing in case.
a. Command area of the appropriation area/command area of the entire system.

43

TABLE 3.2   Bureaucratic Irrigation Systems: Cases Coded

Country	System name	System type	Command area (in hectares)	Major crop	Documentation
India	Kottapalle	Complex	500/—[a]	Rice	Wade (1985 & 1988a)
India	Sananeri	Complex	173/1,172	Rice	Meinzen-Dick (1984)
India	Dhabi Minor Watercourse	Complex	21/—	Other grains	Gustafson & Reidinger (1971) Reidinger (1974 & 1980) Vander Velde (1971 & 1980)
India	Area Two Watercourse	Complex	33/229,000	Other grains	Bottrall (1981)
Indonesia	Area Three Watercourse	Complex	115/33,000	Rice	Bottrall (1981)
Iraq	El Mujarilin	Complex	307/208,820	Other grains	Fernea (1970)
Laos	Nam Tan Watercourse	Complex	100/2,046	Rice	Coward (1980b)
Pakistan	Dakh Branch Watercourse	Complex	152/—	Other grains	Mirza (1975)
Pakistan	Gondalpur Watercourse	Complex	200/628,000	Rice	Merrey & Wolf (1986)
Pakistan	Punjab Watercourse	Complex	96/—	Rice	Lowdermilk, Clyma & Early (1975)
Pakistan	Area One Watercourse	Complex	50/628,000	Other grains	Bottrall (1981)
Thailand	Kaset Samakee	Complex	28/12,000	Rice	Gillespie (1975)
Thailand	Amphoe Choke Chai	Complex	125/12,000	Rice	Gillespie (1975)
Taiwan	Area Four Watercourse	Complex	150/67,670	Rice	Bottrall (1981)

$N = 14$
— = Missing in case.
a. Command area of the appropriation area/command area of the entire system.

44

TABLE 3.3   Other Irrigation Systems: Cases Coded

Country	System name	System type	Command area (in hectares)	Major crop	Documentation
Peru	Hanan Sayoc	Simple	—	Other grains	Mitchell (1976 & 1977)
Peru	Lurin Sayoc 1	Simple	—	Other grains	Mitchell (1976 & 1977)
Peru	Lurin Sayoc 2	Simple	—	Other grains	Mitchell (1976 & 1977)
Mexico	Diaz Ordaz Tramo	Complex	2/150[a]	Other grains	Downing (1974)

NOTE: The production resource of Lurin Sayoc 1 is governed by barriowide rural political officials. The production resources of the other three cases are governed by municipal governments.

$N = 4$

— = Missing in case.

a. Command area of the appropriation area/command area of the entire system.

45

Twenty-nine of the cases are simple systems; eighteen are complex. Forty-one of the forty-seven cases are located in Asia. The sizes of the systems range from 3 hectares (Cadchog, a community system) to 628,000 hectares (Area One, a bureaucratic system). The major irrigated crop in most of the systems is rice.

A majority of the case studies available are on well-managed community systems, so special efforts were taken to locate and code cases on poorly managed community systems. Most of the case studies on bureaucratic systems lack information about specific appropriation areas. This is why only fourteen bureaucratic cases are coded as compared to twenty-nine community cases. I coded all the cases, and another member of the research project reviewed the coding of each case. Disagreements on the proper coding were discussed and resolved in regular meetings.

Because the original forms were designed to code cases about different kinds of common-pool resources, the wording of some of the questions presented in subsequent chapters has been slightly changed from the original versions to fit the present context. The meaning of the questions, however, remains unchanged.

The values for some variables have also been changed. Some variables in the original coding forms consist of four or five values. In this study, because of the limited number of cases available, the number of values for some variables is reduced to two. In the original coding form, for instance, the variable about the supply of water has five values: (1) extreme shortage, (2) moderate shortage, (3) apparent balance, (4) moderate abundance, and (5) significant abundance. In this study, values 1 and 2 are coded as *inadequate*, and 3, 4, and 5 are coded as *adequate*. The coded values of individual variables for each case are reported in the subsequent chapters to give readers an opportunity to check the validity of the coding.

The basic unit of analysis in this study is the *time slice* during which the actions of participants are relatively consistent and the contextual attributes are relatively stable. Contextual attributes are considered relatively stable if the rules governing the appropriation resource, the community of appropriators, and the physical characteristics of the resource are the same throughout the period. When any of these attributes changes, a new time slice begins, and a new operational-level form is coded. Other new forms are also coded to reflect the corresponding changes in contextual attributes. For example, if new operational rules are employed, a new operational rule form will be used to code the new arrangements. There will therefore be more than one case for each irrigation system if more than one time slice is reported in the documents.

The reader may notice that some of the cases bear similar names, such as San Antonio 1 and San Antonio 2, Tanowong T and Tanowong B, and Lurin Sayoc 1 and Lurin Sayoc 2. San Antonio 1 and San Antonio 2 correspond to two different time slices of an irrigation system built by the National Irrigation Administration of the Philippines in San Antonio. San Antonio 1 stands for the period when a watertender oversaw water allocation; San Antonio 2 stands for the period when the position of watertender no longer existed. Tanowong B refers to a period when the appropriation resource had access to an additional source of water, which Tanowong T did not have. Lurin Sayoc 1 stands for a period when the resource was governed by some barriowide rural political officials, and Lurin Sayoc 2 stands for a period when the resource was governed by municipal officials.

Because these forty-seven cases form the basic evidence of this study, the generalizations derived from the study pertain to what has been reported in the case studies. One cannot ascertain whether this sample is representative of the entire population of irrigation systems in the world, but it provides a wide diversity of experiences from which to analyze institutions and collective action in irrigation systems. No other source of evidence describes in detail the experiences of irrigation systems in such diverse physical, community, and institutional settings. This contrasts with most other studies of irrigation systems, which make generalizations based on the experiences of one or two irrigation systems.[8] The present study combines the advantages of detailed information about individual irrigation systems and a larger number of cases than is available to individual field researchers. An analysis of these forty-seven cases enables us to identify how various collective outcomes are associated with different configurations of physical, community, and institutional attributes of irrigation systems.

CHAPTER FOUR

Collective Outcomes and Physical and Community Attributes

The cooperative efforts of cultivators in investment and water allocation in an irrigation system affect the level of water supply, the degree of rule conformance, maintenance, and the distribution of benefits and costs among the participants. These outcomes are related to one another in a recursive manner. On the one hand, rule conformance and maintenance may act as independent variables affecting the level of water supply. By cooperating to maintain an irrigation system, for instance, cultivators can increase the level of water supply in the system. After they have utilized an appropriation resource for several years and are able to enforce a suitable set of water allocation rules, they can estimate the usual amount of water available and plan the number and types of crops to be cultivated accordingly. On the other hand, the level of water supply may act as an independent variable affecting the degree of rule conformance and maintenance: Cultivators' incentives to cooperate in water allocation and maintenance are affected by the ability of water available from the appropriation resource to meet the requirements of crops.

The collective outcomes in an irrigation system are also related to various physical and community attributes of the system. The extent to which cultivators depend on an irrigation system may affect their incentives to cooperate. After individual cultivators have realized the potential benefits of cooperation, they have to expend resources to organize among themselves and assign responsibilities

49

to undertake actual water allocation and maintenance. The size of the irrigated area, the number of cultivators involved, the distribution of wealth, and the social and cultural differences among cultivators affect their coordination costs and their abilities to develop and sustain institutional arrangements that can solve their problems.

In this chapter, I first discuss collective outcomes as they are found in the cases. The outcomes in these cases occur in a specific pattern, an analysis of which enables one to draw inferences about the relationships of the outcomes to one another. I then examine how various physical and community attributes are associated with different outcomes. Some of these attributes tend to be associated with inferior outcomes. Others affect outcomes either positively or negatively, depending on the configuration of other contextual variables.

Water Supply, Rule Conformance, and Maintenance

The performance of an irrigation system can be measured in various ways. On the technical side, one may measure the marginal productivity of the water used for irrigation or the proportion of water loss through seepage during conveyance. Because time-consuming technical surveys are needed to obtain them, these measurements are absent from most case studies. Notwithstanding this lack of technical information, most cases do report information regarding the relative adequacy of water supply, level of maintenance, and degree of rule conformance among cultivators in an irrigation system. These outcomes can serve as rough measures of the relative performance of an irrigation system.[1]

The evidence regarding these outcomes varies from case to case. In some cases, the author discusses the outcomes specifically; in others, inferences must be drawn from related discussions in the case. The questions used in the coding forms to identify outcomes covered three areas: adequacy, rule conformance, and maintenance.

1. *Adequacy:* At the end of this period, does the amount of water available in the appropriation resource meet the water requirements of the crops in the established fields served by the resource?

Because the principal objective of any irrigation system is to supply water for agriculture, it is important for an irrigation system to have enough water to meet the needs of crops planted by appropriators of the system. The following excerpts are examples of the authors' assessment of the relative adequacy of water supply in some case studies.

Cases in which the water supply is adequate:

Raj Kulo: There have been significant improvements made in the canal, and the amount of water supplied to the command area has increased considerably in the past 25 years, but there has been little increase in the area that is irrigated. . . . Whereas they once had to use a rotation system of distribution and go out to irrigate at night, now the water flows continuously to all fields, and much of the time, excess water is diverted to a drain. (Martin and Yoder 1983a, 24–25)

Cadchog: Our informants claim that their system's water supply is always sufficient for the irrigation needs. They explain that they can always obtain adequate irrigation in spite of their being situated at the downstream portion of the creek because the upstream dams do not divert all the creek's water. (de los Reyes et al. 1980a, 85)

Cases in which the water supply is inadequate:

Tanowong T: Over the years the inadequacy of the original irrigation sources became more and more of a problem for three reasons: (1) the expansion and increase in the number of terraces on the original site, (2) the construction of new terraces along and below the irrigation ditch which necessarily diverted water permanently, and (3) the denuding through careless cutting and frequent fires of the pine forest of the mountains in the environs of the streams which served as the source of irrigation water. Gradually, therefore, more and more of the original rice terraces, particularly those located in the lowest tiers, were not adequately watered and thus became increasingly unproductive, leading to their conversion to the growing of sweet potatoes. (Bacdayan 1980, 177)

Dhabi Minor: The total amount of water available to each farmer is severely limited; each season most farmers can irrigate only about one-third of their land included in the canal service area. . . . Canal-irrigation supplies have at least three types of uncertainty or unreliability with which the farmer must contend: the timing of water supplies during the season, the quantity of water to be received at various times during the season, and in total, and the timing and quantity of water received at various locations. This uncertainty is in direct contrast to the farmer's allocated turn to receive water. (Reidinger 1980, 269, 281-83)

As these examples show, elements such as the quantity of water available from the source, the number and types of crops cultivated, and the timeliness and reliability of the water supply schedules jointly determine whether an adequate supply of water is available in an irrigation system. Twenty-one cases, or 45 percent, indicate an adequate supply of water (see Figure 4.1).

The coding on the level of water supply is affected by how an irrigation system is initially defined. In some cases, irrigators in an appropriation resource have access to water derived from a separate water system. This separate system, when combined with the original irrigation system, may provide sufficient water for cultivation. The water supply from the original system is considered

FIGURE 4.1   Positive Outcomes on Three Measures of
                  Irrigation System Performance (all 47 cases)

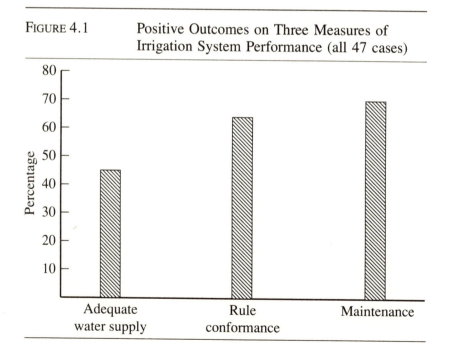

inadequate, however, if the original system alone does not provide sufficient water for cultivation. In Sananeri Tank, for example, farmers rely primarily on water from the surface irrigation system. In times of water scarcity, they have access to groundwater through pumpsets that are owned and operated privately by individual farmers. Because these pumpsets are operated and governed separately from the surface irrigation system, they are not considered parts of the surface system. The level of water supply from Sananeri Tank is considered inadequate because farmers cannot have enough water for cultivation without supplements from the ground. Besides Sananeri Tank, there are two other cases in the sample—Char Hazar and Amphoe Choke Chai—where farmers are able to get sufficient water for cultivation from water sources other than the original irrigation systems. In these two cases, the levels of water supplies from the original systems are still considered inadequate. The presence or absence of alternative sources of water affects farmers' incentives to cooperate with one another. The relationships between alternative water sources and collective action are discussed later in this chapter.

2. *Rule conformance:* Do most irrigators follow the local operational-level rules-in-use related to this appropriation resource in years when there is no extreme shortage?

Operational rules are important means of coordinating water allocation and maintenance among appropriators. The willingness of most appropriators regularly to follow them reflects the viability of these rules as coordinating devices. The following are examples of evidence for these outcomes in some case studies.

Cases in which most appropriators follow operational-level rules in use:

Pinagbayanan: In the dry season, the members paid their obligations in cash rather than in kind.... The total collection amounted to P5971.... The association was then able to repay its P10000 loan from the rural bank of Pila. This was a remarkable achievement and it attests to the members' concern for living up to their commitments in a cooperative way. (Cruz 1975, 255)

Deh Salm: During a three-month field season, I did not encounter any case of one individual infringing upon the rights of another by keeping the water flow beyond the time limit of his share. I was told it never happened, and I deduce that it is in every individual's interests to keep the system working smoothly. (Spooner 1974, 44)

Cases in which substantial numbers of appropriators fail to follow operational-level rules in use:

Char Hazar: The indigenous farmers' irrigation organization has begun to deteriorate. Traditional rules and regulations are no longer followed, and maintenance tasks are not performed as well as in previous years. (Fowler 1986, 59)

San Antonio 1: The association officials instructed the watertender to rotate the distribution of water. . . . This arrangement was observed only for 2 days, however. All of the farmers disregarded the schedule after one of the upstream farmers diverted some water on the second night. (de los Reyes et al. 1980b, 53)

Thirty of the cases, or 64 percent, indicate that most appropriators follow operational rules in use.

3. *Maintenance:* At the end of this period, is the appropriation resource well maintained?

Besides the care taken by the appropriators, a wide diversity of elements affects the maintenance of an irrigation system. These may include the initial construction and physical environment of the system and the financial and technological capabilities of the irrigators. One must take these elements into account when determining whether an appropriation resource is well maintained. Examples of evidence for this outcome from specific case studies follow.

Cases in which the appropriation resource is well maintained:

Na Pae: Because the canal bank is strong and built firmly with rocks, Na Pae members seldom have maintenance problems. However, in some sandy areas where the bank easily slips, there has been trouble before the bank was repaired with concrete. . . . Since the irrigation system is small and has never been threatened by a natural catastrophe, it requires relatively little work; one good maintenance effort a year is able to keep the system in good working condition. . . . A few locations along the canal have persistent problems of sand slides and leaking, but when people repair the canal bank with cement the problem is permanently solved. (Tan-kim-yong 1983, 209, 217–18)

Thulo Kulo: Over the years, improvements have been made to the main canal, significantly increasing the total flow in the system. . . . Improvements in the canal have been made on an almost annual basis. This has resulted in increased discharge from a mere trickle in 1932 when the canal irrigated only a few small plots, to a maximum discharge of 180 liters/second in 1982. The average discharge, measured in the main canal on a twice daily basis over the 1982 rice season, was 160 liters/second. (Martin and Yoder 1986, 10)

Cases in which the appropriation resource is poorly maintained:

Gondalpur: At the time of the study (1976–1977), the level of maintenance of all the branches on the watercourse was extremely poor. . . . For some years after the installation of the tubewell, there was no perceived shortage of water. According to informants this led to a decrease in maintenance efforts, atrophying the already weak sanctions enforcing participation in watercourse cleaning. . . . The watercourse on all branches was choked with grass, bushes, and trees; leaked through rat holes, thin banks, and at junctions; and water remained standing in many low sections after irrigation. (Merrey and Wolf 1986, 35)

San Antonio 1: The system's main canal had deteriorated. For instance, the areas of the canal where the riprapped portions begin and end had become wider and deeper; hence, the water would collect into some sort of a pool along these portions of the canal and consequently, the water that flowed after each riprapped portion decreased. Also, at about 3/4 kilometer away from the dam, the water fell into a drop. Here the water formed a sort of basin, and from this point only a portion of the water continued toward the downstream portion of the canal. . . . At the start of the third crop season, the canal conditions had worsened and the volume of water reaching the downstream area reduced to a trickle. (de los Reyes et al. 1980b, 52–53)

Thirty-three of the cases, or 70 percent, indicate that the appropriation resource is well maintained.

Patterns of Outcomes.   Level of water supply, degree of rule conformance, and maintenance are closely related to one another. An adequate supply of water encourages a high degree of rule conformance and maintenance, and vice versa. Within the sample of cases, these three outcomes are associated with one another in a pattern that resembles a Guttman scale. In a Guttman scale, the component items can be arranged in a systematic and cumulative fashion so that there is "a continuum that indicates varying degree of the underlying dimension" (Nachmias and Nachmias 1987, 475). By employing this property of Guttman scales, one can predict the sequence of collective outcomes generated in irrigation systems.

The Guttman scale as shown in Table 4.1 can be interpreted in two complementary ways. One interpretation holds that the outcomes are arranged cumulatively along a continuum of increasing degree of difficulty. Some outcomes are more difficult to attain than others: A case that is characterized by a difficult outcome will usually be characterized by a less difficult outcome, but not vice versa. Within the sample of cases, an adequate supply of water is the most difficult to attain. The degree of difficulty is followed by a high degree of rule conformance and good maintenance. Forty-

TABLE 4.1	Three Measures of Irrigation System Performance Arranged on a Guttman Scale		
Good maintenance	Rule conformance	Adequate water supply	Number of cases
yes	yes	yes	21
yes	yes	no	8
yes	no	no	4
no	no	no	13
no	yes	no	1
		Total =	47

CR (coefficient of reproducibility) $= 1 - 1/47 \times 3 = 0.99$

six of forty-seven cases conform perfectly to the scalable pattern. The coefficient of reproducibility, which measures the degree of conformity to a perfect scalable pattern, is 99 percent.

If the pattern is perfect, an irrigation system with a high degree of rule conformance will also be well maintained; a case with an adequate supply of water will have both a high degree of rule conformance and good maintenance. All twenty-one cases with adequate supplies of water are characterized by both high degrees of rule conformance and good maintenance (see Table 4.2). Among these cases, the three outcomes tend to reinforce one another. On the one hand, an adequate supply of water encourages appropriators to cooperate in water allocation and maintenance. On the other hand, high levels of rule conformance and good maintenance enable irrigators to further develop and preserve their water supplies.

The Guttman scale also shows that even in cases characterized by good maintenance and a high degree of rule conformance, the supply of water may still be inadequate. Of the twenty-nine cases with a high degree of rule conformance and good maintenance,

TABLE 4.2	Rule Conformance and Maintenance by Adequacy of Water Supply		
	Adequate water supply	Inadequate water supply	Total
Cases rated positive in both rule conformance and maintenance	100% 21	31% 8	29
Cases rated negative in either rule conformance or maintenance or both	0% 0	69% 18	18
Total	100% 21	100% 26	47

Percentage difference = 69%
Chi-square with continuity correction factor = 20.7
D.F. = 1   P < 0.0001

eight are characterized by an inadequate supply of water (see Table 4.2). These eight cases show that even if appropriators cooperate in rule enforcement and maintenance, an appropriation resource may still have an inadequate supply of water. Water scarcity may be a result of constraining factors other than the appropriators' failure to enforce rules and maintain appropriation resources.

Nayband, for example, is an oasis on the Iranian Plateau that has plenty of land but a limited supply of water from nearby springs. Water inadequacy is an environmental constraint beyond irrigators' immediate control. Other examples are Kottapalle, Sananeri, and Nam Tan Watercourse, all of which are located in complex bureaucratic irrigation systems. In these systems, the amount of water available to an appropriation area is affected by such factors as the location of the area within the larger system and decisions by officials responsible for releasing water from the main canal to the area. These factors are beyond the immediate control of the irrigators withdrawing water from the appropriation area. In some situations the level of water supply may not be a pertinent indicator of the success or failure of collective action by appropriators. The adequacy of water supply can therefore be treated as a contextual attribute that affects the structure of incentives facing appropriators.

Another interpretation of the Guttman scale in Table 4.1 holds that problems in irrigation systems are arranged cumulatively along a continuum of increasing severity. If a more severe problem is present, the less severe ones are usually also present, but not vice versa. In other words, problems in irrigation systems usually appear in a specific sequence: First, the water supply is scarce or poorly matched to the standing crops; then, more and more irrigators fail to follow allocation and maintenance rules; and finally, the maintenance of the appropriation resource begins to deteriorate. If an appropriation resource has an adequate supply of water, it will seldom have problems in rule conformance or maintenance. If problems do occur in an appropriation resource, water scarcity is usually the first to emerge; if additional problems follow, they will be in the areas of rule conformance and maintenance.

Within the sample, all eighteen cases with problems in rule conformance or maintenance are also characterized by an inadequate supply of water (see Table 4.2). Although one cannot infer from the pattern that all problems in rule conformance and maintenance are caused directly by water scarcity in an appropriation resource, water scarcity appears to be a major cause of these problems in many individual cases. In these cases, it is also the first problem to emerge, subsequently inducing conflict among appropriators and affecting their willingness to follow allocation and maintenance schedules. A case in point is San Antonio of the Philippines, where water shortage created a high level of conflict among appropriators and eventually reduced the ability of the irrigation association to enforce its allocation and maintenance rules. Another example is Mauraro of the Philippines, where farmers routinely pierced the canal embankment to increase the flow of water to their fields, thus making it more difficult to maintain the irrigation system.

According to the Guttman scale, cases with an inadequate supply of water may or may not have problems in maintenance and rule conformance. In eight of the twenty-six cases with inadequate supplies of water, most irrigators still follow operational rules in use and maintain their appropriation resources well (see Table 4.2). This shows that irrigators may be able to overcome obstacles for collective action created by water inadequacies. Indeed, Wade (1988a) found that the more scarce and uncertain the water supply is in south India, the greater is the likelihood that a community of cultivators will develop collective arrangements to manage their watercourse. One of the reasons why cultivators in south India will organize in a situation of water scarcity is that they can influence the amount of water available in their village by concerted actions such as bribing officials and intimidating upstream stealers. Water scarcity, in this particular case, acts as an additional incentive for cultivators to get organized.

On the other hand, if farmers do not have much chance of increasing their water supply, an inadequate supply of water may reduce their incentives to organize for allocation and maintenance. This is probably the case in the Philippine irrigation system to which Wickham and Valera (1979) refer when they argue that in order to induce farmers to cooperate in managing their watercourses, an effective systemwide management program is a prerequisite. Their argument implies that if farmers do not have much influence on the amount of water that flows into their watercourse, they have less incentive to cooperate than if they have a reliable and adequate flow of water into their watercourse in the first place.

However the Guttman scale is interpreted, the pattern of outcomes identified in the sample indicates that the chance of having a high degree of rule conformance and good maintenance is smaller in systems with inadequate supplies of water than in those with adequate supplies. Although irrigators' failure to organize water allocation and investment may adversely affect the level of water supply in an appropriation resource, an inadequate supply of water may, conversely, hinder collective action among irrigators.[2]

Distribution of Benefits and Costs

An additional performance indicator of an irrigation system is the distribution of benefits and costs among its appropriators. One question was used on the coding forms to identify this outcome:

- *Disadvantaged:* Are there any appropriators who have been consistently disadvantaged in this period?

Few irrigation systems benefit all irrigators equally because different irrigators may cultivate different amounts of land. Instead of asking whether every irrigator gets an equal amount of water, one may inquire whether some groups of irrigators consistently get a disproportionately smaller amount of water to cultivate their crops or provide a disproportionately larger number of investments than others. Fourteen of thirty-seven cases, or 38 percent, indicate that some appropriators have been consistently disadvantaged in this sense (see Table 4.3).

A group of appropriators may be consistently disadvantaged in two ways. One involves a common problem in most canal irrigation systems where headenders have a natural advantage over tailenders in their access to water. If the supply of water in the appropriation resource is limited and most appropriators fail to follow allocation procedures and maintain the water delivery facilities, tailenders are likely to get less water than headenders. Ten out of fourteen cases that report the presence of a disadvantaged group of appropriators appear to fall into this category (see Table 4.3). All ten cases are characterized by an inadequate supply of water and problems in rule conformance or poor maintenance. Lowdermilk, Clyma, and

TABLE 4.3   Rule Conformance and Maintenance by
                   Disadvantaged Groups

	Without disadvantaged groups	With disadvantaged groups	Total
Cases rated positive in both rule conformance and maintenance	87% 20	29% 4	24
Cases rated negative in either rule conformance or maintenance or both	13% 3	71% 10	13
Total	100% 23	100% 14	37

Percentage difference = 58%
Chi-square with continuity correction factor = 10.6
D.F. = 1 $P < 0.01$

Early, for example, describe the situation in Punjab Watercourse as follows:

> Given the present system with losses resulting from seepage, dead storage, countless leaks and spills along watercourses at improper elevation along unlevel fields, there is actually a built-in mechanism creating mini-type economic dualism between watercourse users located at the head and tail positions. The farmers who own land at the tail are always at a disadvantage for canal water. (1975, 27)

If the problems with water supply, rule conformance, and maintenance in these watercourses are alleviated, the position of the tailenders will improve considerably.

In certain irrigation systems, some irrigators are consistently disadvantaged as a result of institutional arrangements instead of problems in water supply, rule conformance, or maintenance. Four of the fourteen cases that report the presence of a disadvantaged group of appropriators fall into this category (see Table 4.3). One example is Kheri in Tanganyika—now Tanzania—where appropriators were divided into four groups. Two hereditary groups had absolute priority to irrigation water. The other two groups had to purchase water rights from the two hereditary groups. One of the hereditary groups was responsible for managing the irrigation system and, at the same time, acted as "the political rulers of the village with powers to issue orders and constitute themselves a court of law" (Gray 1963, 164). It appears that water privileges and political powers in the village reinforced each other.

Another example is Felderin in the Swiss Alps, where water allocation is based on specific time slots owned by different individuals. In this system, "there are gross inequalities in amount of water available per unit of land" (Netting 1974, 73). Arguing that these inequalities are results of subdivisions of properties through inheritance and sale, which at times led to exchanges and reapportionment of water rights, Netting writes:

> A rationalized system of water sharing is resisted by those who derive advantage from the current arrangement. Convenient watering periods during the day are valued, and owners are reluctant to accept other times. Though everyone recognizes that some unfairness of distribution is perpetuated by the existing system, large owners claim (1) that their water is as much a possession as the land and is subject to similar inequalities in tenure, (2) that any reorganization would be dreadfully complicated, and (3) such a project would inevitably arouse suspicion and animosity in all concerned. (1974, 73)

Some commentators are concerned that indigenous organizations tend to perpetuate inequalities among farmers. They argue that the decision-making processes in many of these communities are dominated by the local elite. The poor and less influential farmers are usually disadvantaged in their access to common-pool resources in the communities. An examination of the sample, however, does not support this contention. Only four out of twenty-three community irrigation systems in the sample are characterized by institutional arrangements that are specifically designed to favor one group of irrigators over another.

Most bureaucratic irrigation systems are designed to supply water to whoever cultivates crops in a particular area. No one is supposed to be discriminated against by design. However, problems in water supplies, rule conformance, and maintenance in many of these systems put some of their irrigators, especially the poorer ones who cultivate only lands located in the tail portion of a watercourse, into consistently disadvantaged positions.

Dependence on an Appropriation Resource

Irrigators frequently have to invest their private resources or forgo some immediate, short-term benefits in order to follow and enforce allocation and maintenance rules. Their incentives to contribute to these investments may be affected by their degree of dependence on the resource. Two questions in our coding forms indicate farmers' relative dependence on an appropriation resource:

- *Family income*: How dependent are most of the appropriators on this appropriation resource as a major source of family income? (That is, do most of the appropriators' family incomes come directly from cultivating crops irrigated by the resource?)

- *Alternatives*: Do most appropriators have access to an alternative source of water for irrigation?

In about half of the cases, most irrigators derive most of their income directly from cultivating crops irrigated by the resource. In the other half, most irrigators have other sources of income. These alternative sources of income include cultivating crops irrigated by other water sources, raising livestock, and working at jobs outside the agricultural sector. In 40 percent of the cases, most appropriators have access to alternative sources of water for irrigation. These sources include groundwater basins and other surface resources.

Within the sample, neither the availability of alternative sources of income nor the availability of alternative sources of water appear to be related to rule conformance and maintenance (see Tables 4.4 and 4.5). These results suggest either (1) that farmers' degree of dependence on an appropriation resource does not affect rule conformance and maintenance or (2) that their effects could be either positive or negative depending on other contextual factors.

Some of the case studies illustrate how appropriators' degree of dependence on an appropriation resource affects their incentives to

TABLE 4.4   Rule Conformance and Maintenance by
Family Income

	Most of family income from resource	Half or less of family income from resource	Total
Cases rated positive in both rule conformance and maintenance	68% 15	58% 11	26
Cases rated negative in either rule conformance or maintenance or both	32% 7	42% 8	15
Total	100% 22	100% 19	41

Percentage difference = 10%
Chi-square with continuity correction factor = 0.13
D.F. = 1 $P > 0.5$

cooperate under various circumstances. Near Char Hazar in Nepal, for example, a new irrigation system was constructed from which water leaked to the Char Hazar system. Farmers in Char Hazar gradually became dependent on the leakage water. Now, most farmers are no longer willing to follow traditional rules and maintenance schedules. The experiences of Char Hazar show that if farmers believe they can get enough water from an alternative source without any extra effort, they will have less incentive to cooperate in water allocation and maintenance.

The presence or absence of alternative water sources also affects farmers' strategies in various water zones within the Lam Pra Plerng Irrigation Project—a bureaucratic irrigation system—in Thailand (Gillespie 1975). Zone One (Kaset Samakee), located at the head end of the system, had a reliable supply of water and no alternative source of water. Most farmers in the zone followed water allocation schedules and participated in maintaining ditches. This zone contrasts with the other six zones of the system that were plagued by problems of water allocation and maintenance. In two

TABLE 4.5   Rule Conformance and Maintenance by
              Alternative Water Supply

	With alternative water supply	Without alternative water supply	Total
Cases rated positive in both rule conformance and maintenance	53% 9	58% 14	23
Cases rated negative in either rule conformance or maintenance or both	47% 8	42% 10	18
Total	100% 17	100% 24	41

Percentage difference = 5%
Chi-square with continuity correction factor = 0.001
D.F. = 1   P > 0.5

of these zones, located at the tail end of the system, farmers also depended on water from the system, but the supply reaching the zones was both erratic and limited. In the other four zones, located at the middle part of the irrigation system, farmers usually received sufficient water from the natural flooding of rivers and did not depend entirely on water from the system. In the words of Gillespie:

> With the introduction of the new irrigation system, Zone One received a dependable supply of water. The farmers took advantage of this and began planting paddy. The higher incidence in Zone One of farmers cleaning their farm ditches when water is scarce may therefore be related to their comparative dependence on irrigation water, since they have no alternative sources. Moreover, the generally porous nature of the soil necessitates the distribution of water as quickly as possible, for the longer it takes to distribute the water the more is lost by seepage. If water can not flow easily through the ditches because of silt or weeds, the farmers are compelled to keep them clean. (1975, 7)

It appears that the absence of alternative water sources, in combination with a reliable supply of water and extra efforts required to keep the water flowing in the channel, created powerful incentives for farmers in Zone One to cooperate in water allocation and maintenance. As shown by the two tail-end zones in the system, farmers dependent on water from an appropriation area may not have incentives to organize in water allocation and maintenance if they have a highly unreliable supply of water in the first place. The four zones located in the middle part of the system show that if farmers have access to an inexpensive, alternative source of water, they may not have much incentive to cooperate in governing and maintaining their appropriation resource.

Sananeri in India is another case in which the availability of an alternative source, well water, appears to have both positive and negative effects on farmers' incentives to govern their surface water system (Meinzen-Dick 1984, 62–72). On the one hand, irrigators' potential access to well water encourages them to cooperate in maintaining and securing more water for their surface system. Because it is relatively expensive to extract water from the ground, irrigators have incentives to keep the field channels clean and in good repair in order to move the high-value well water through these channels as efficiently as possible. The availability of well water also helps to ease tension among appropriators when the water supply in the surface system is scarce. On the other hand, groundwater also creates a potential conflict of interest among irrigators. Well owners want the irrigators' association to use its resources to lobby for more frequent water issues from government officials (in order to replenish the water basin). Irrigators without wells may, however, be unwilling to share those expenses from which they receive little direct benefit. In spite of this negative effect, an alternative water source, on balance, facilitates cultivators' cooperation in governing Sananeri.

The experiences of these cases suggest that the degree to which farmers depend on an irrigation system does not directly affect their collective action in irrigation systems. A high degree of dependence on an irrigation system may increase or decrease farmers' incentives for cooperation, depending on the configuration of other contextual factors.

Irrigated Area and Number of Irrigators

Although the information-gathering, communication, decision-making, and monitoring costs for governing a resource tend to increase as the size of the resource increases, large resources are not doomed to failure. Depending on the geographical and hydraulic environments, it may be more economical to develop irrigation systems and watercourses that serve large numbers of fields and irrigators. Appropriate institutional arrangements can help to solve coordination problems in large irrigation systems and appropriation resources.

Three questions in our coding forms are used to assess this dimension of a case:

- *Appropriation size*: At the end of this period, how many hectares of fields are irrigated by the appropriation resource?

- *Number*: At the end of this period, what is the number of appropriators utilizing the appropriation resource?

- *System size*: How many hectares of fields are irrigated by the entire irrigation system, including production, distribution, and appropriation resources?

As discussed in Chapter 3, a simple irrigation system consists of only one appropriation resource. For this kind of case, *appropriation size* and *system size* are the same. A complex irrigation system consists of multiple appropriation resources. For this kind of case, *appropriation size* and *system size* are different.

Within the entire sample of cases, the amount of land and number of irrigators served by an appropriation resource fail to show a strong relationship with the level of rule conformance and maintenance in the resource (Table 4.6). Within the sample of complex cases, the total amount of land served by a system also fails to show a significant relationship with the degree of rule conformance and maintenance in its appropriation resource (Table 4.7). Even though coordination costs tend to increase as the numbers of fields and appropriators increase, appropriators are still capable of overcoming

TABLE 4.6   Logit Estimates of the Relationship between Rule Conformance and Maintenance and Measures of Appropriation Resources

Variables	Coefficient	t-score
Dependent:		
Rule conformance and maintenance[a]		
Independent:		
Hectares of field irrigated by the appropriation resource	0.0036	1.70[b]
Number of appropriators utilizing the appropriation resource	−0.0006	0.54[c]

$N = 37$
R-square $= 0.096$
Adjusted R-square $= 0.04$
a. Positive in both rule conformance and maintenance $= 1$;
   negative in either rule conformance or maintenance or both $= 0$.
b. $P > 0.05$
c. $P > 0.5$

these costs and successfully organizing their water allocation and maintenance activities in irrigation systems and appropriation areas of substantial sizes. Kottapalle in India, for example, is an appropriation area that serves about 500 hectares of land and 800 irrigators. Another example is El Mujarilin in Iraq, which serves more than 300 hectares of land; the irrigation system where El Mujarilin is located serves more than 200,000 hectares of land.

Social and Cultural Divisions

If a community of irrigators is divided by ethnic, cultural, clan, racial, caste, or other social differences that inhibit communication, the costs of organizing collective action within the community will be higher than in those without divisions. One question was used to identify this attribute:

TABLE 4.7   Logit Estimates of the Relationship between Rule Conformance and Maintenance and System Size among Complex Systems

Variables	Coefficient	t-score
Dependent:		
Rule conformance and maintenance[a]		
Independent:		
Hectares of field irrigated by the entire irrigation system	-0.0000	-0.93[b]

$N = 12$
R-square $= 0.38$
Adjusted R-square $= 0.31$
a. Positive in both rule conformance and maintenance $= 1$;
   negative in either rule conformance or maintenance or both $= 0$.
b. $P > 0.1$

- *Cleavages*: Are there any ethnic, cultural, clan, racial, caste, or other differences among the appropriators that may affect their capacities to communicate with one another effectively?

In the sample, seven cases (two community and five bureaucratic cases) are reported to have divisions among irrigators that inhibit their communication with one another. The two community cases are characterized by both a high degree of rule conformance and good maintenance; the five bureaucratic cases are characterized by both a low degree of rule conformance and poor maintenance.

Chiangmai is an appropriation resource located in a complex community irrigation system in Thailand. Farmers within the Chiangmai village are divided into two major factions. Although this division has created numerous conflicts among farmers in the village, they are able to cooperate on irrigation matters.

In Deh Salm, a community irrigation system in the Iranian Plateau, six brothers from outside the village purchased water shares of the system and financed the improvement of the *qanat* (tunnel) that diverts water to the appropriation area. After that, they gained rights to cultivate land in the village. Later "some of the brothers . . . sold out, and others or their heirs . . . settled on one side of the village and [became] resident cultivators, performing some, if not all, of their own cultivation" (Spooner 1974, 53). Although these people have become residents in the village, they are not considered members of the community. The division between members and non-members in the village does not, however, undermine the functioning of the system, because its operation and maintenance usually do not require much active involvement and contributions by the irrigators. Unless the *qanat* is damaged by extreme circumstances, the entire irrigation system does not require much maintenance. Water is allocated according to water shares that correspond to specific time slots in a distribution cycle. This allocation arrangement is self-enforcing because every shareholder has incentives to guard his own time slots. As long as the *qanat* requires no major repair, the irrigation system remains viable in spite of the social division in the village.

Five of the bureaucratic cases (Area Two Watercourse, Gondalpur Watercourse, Dakh Branch Watercourse, Dhabi Minor Watercourse, and Punjab Watercourse), all located in either India or Pakistan, are reported to have communication problems due to social divisions among their participants. In many parts of India and Pakistan, farmers are divided into various caste and subcaste groups, which are further subdivided into kinship or brotherhood groups (Merrey and Wolf 1986; Lowdermilk, Clyma, and Early 1975). Although these divisions may not inhibit communication and cooperation among farmers in every irrigation system in the two countries, they do make cooperation among irrigators in these five cases more difficult. All five cases are characterized by inadequate supplies of water, poor maintenance, and low levels of rule conformance among irrigators.

The cleavages among *biradaris* (kinship groups) in some Pakistani communities are reinforced by the cultural concept of *izzart* (Merrey and Wolf 1986). *Izzart* may be translated variously as *honor, esteem,* or *face.* People regard the *izzart* game as zero-sum in nature, meaning that one acquires *izzart* only at someone else's expense; "the success of one person is a threat to all the other players, a characteristic that generates competition and jealousy"

(Merrey and Wolf 1986, 38). The concept of *izzart* may be applied at both an individual and a group level. Many disputes in the Gondalpur, for example, stem from *izzart* games among *biradaris*. There is a feeling that the *izzart* of a *biradari* must be protected. If the *izzart* of a member is hurt by a person from a different *biradari*, other close kinsmen are obliged to unite against the offender. The concern for *izzart* hinders cooperation among irrigators:

> Men oppose or support decisions and programs based on their perceptions of their competitors' position. For example, even though all farmers suffered the exactions of a corrupt tubewell operator, they did nothing because, informants explained, if one man or group proposed petitioning for his removal, others would oppose. This would be done not out of love for the tubewell operator but to prevent the others from gaining some advantage from the issue or to pursue some long-standing grudge. This can be carried further: the non-cooperative behavior of [a *biradari*] on branch A during the watercourse reconstruction was interpreted by informants as based on a desire to prevent others from benefiting—even if it means foregoing their own potential benefits. (Merrey and Wolf 1986, 39)

Before the British rule, these kinds of conflict were mostly avoided by organizing social activities in small groups. Ancestors of the Gondalpur farmers were cattle herders and part-time farmers. The society was characterized by relative mobility of individuals and families; people moved around in small groups. This dispersion of the population helped to avoid conflict that would otherwise be rampant. After the irrigation system was constructed under the British rule, people began to settle down and became full-time farmers. The irrigation system creates situations that require cooperation among many farmers. Traditional cleavages among them, however, become obstacles to their cooperation.

In some other cases, institutional arrangements are developed to mitigate potential conflict in larger groups. In Punjab Watercourse, for instance, informal water turn schedules within the watercourse are adjusted to follow family lines in order to minimize disputes (Lowdermilk, Clyma, and Early 1975, 40). Other cooperative ventures such as the ownership and operation of *jalars*—persian wheels used for lifting water from a shallow depth—within the watercourse are usually organized among kinship members. Although kinship groups occasionally exchange water turns, their cleavages remain a potential obstacle for cooperative actions on a larger scale.

Distribution of Wealth

The distribution of wealth among irrigators may affect their collective action in water allocation and maintenance. One question was used on our forms to identify this attribute:

- *Income variance*: What is the variance of the average annual family income across families among appropriators?

In the sample of cases, a low variance of the average annual family income among irrigators tends to be associated with a high degree of rule conformance and good maintenance. A higher percentage of the cases with low income variance is characterized by both a high degree of rule conformance and good maintenance than of the cases with high income variance (Table 4.8).

TABLE 4.8   Rule Conformance and Maintenance by
                   Income Variance

	Low income variance	Moderate income variance	High income variance	Total
Cases rated positive in both rule conformance and maintenance	89% 8	75% 9	17% 1	18
Cases rated negative in either rule conformance or maintenance or both	11% 1	25% 3	83% 5	9
Total	100% 9	100% 12	100% 6	27

Chi-square = 9.1
D.F. = 2   P < 0.05

This result, however, has to be treated cautiously. Few cases in the sample specifically discuss how the distribution of wealth affects cooperation among irrigators. Only twenty-seven of the cases provide enough information to estimate roughly the degree of income variance among irrigators. Because of the limited number of cases available and the sketchy nature of the information, the test of relationship between income variance and rule conformance and maintenance is merely suggestive. Further, the relationship probably cannot be attributed to income variance alone because the five cases with both high income variance and problems in rule conformance and maintenance are all bureaucratic cases that are also characterized by other features unfavorable to collective action. High income variance is probably one factor among others that create collective-action problems in these systems. The limited number of cases available, however, prevents one from assessing the relative importance of these factors.

Locational Differences

Locational differences are a major source of collective-action problems in many irrigation systems. As discussed earlier, locational differences in conjunction with an inadequate supply of water, a low degree of rule conformance, and poor maintenance may put some irrigators in consistently disadvantaged positions. The influence of locational differences on farmers' collective action in irrigation is documented in some cases.

Depending on how plots are distributed along the main canal in a watercourse, irrigators face different incentives for cooperation. Mirza and Merrey (1979), in a study of ten watercourses in Pakistan, find that a watercourse is more likely to be well maintained if power and influence are concentrated at the tail or at the tail and middle of the watercourse. This is because the powerful and influential people have incentives to help organize water allocation and maintenance activities in the watercourse so that sufficient water can reach their fields located in the middle and tail portions of the watercourse.

In Kottapalle in India, the fields of any one household tend to be scattered about the appropriation area (Wade 1988a). This pattern of scattered plots is especially prevalent among farmers with large holdings. Although the pattern is partly a result of partible inheritance, it is also a way to minimize risk. Land in the area is extremely variable "in soil type, sub-surface drainage, slope, susceptibility to flash floods, and micro-climate" (Wade 1988a, 47). The possession of fields scattered throughout the village ensures against the loss of all crops at the same time. Scattered holdings also enhance collective action in irrigation: Farmers with land in the head end of the canal may have plots near the tail end; this creates a common interest in rules that facilitate water allocation throughout the entire appropriation resource.

In some community irrigation systems, irrigators' associations adopt rules specifically to ensure that members have fields in the head, middle, and tail portions of the major canals. For instance, within each watercourse in Zanjera Danum in the Philippines, land along a lateral canal is divided into several blocks perpendicular to the source of water (Coward 1979). The blocks thus represent differential distances from the water source: Some are near the head end of the canal, some near the tail end. Each of the blocks is further divided into several parcels. Each share in the irrigation system is tied to one parcel in each block, so that each share-holder has to cultivate parcels at various distances from the water source. This arrangement motivates all irrigators to help deliver water throughout the entire watercourse. When there is not enough water to irrigate an entire watercourse, decisions can be made to discontinue irrigating some of the blocks. In this way, the burden of water scarcity is borne by all irrigators proportionally.

Further, in Zanjera Danum, one or more parcels at the tail-end portion of each watercourse are reserved for the irrigation leaders, who are allowed to farm these parcels as a compensation for their services to the irrigators. This arrangement encourages the irrigation leaders to work diligently to deliver water efficiently from the head to the tail of the watercourse.

Summary

The pattern of outcomes found in the sample of cases suggests that an inadequate supply of water is a more common problem in irrigation than a low degree of rule conformance and poor maintenance. Although the failure to organize water allocation and maintenance causes water scarcity in some cases, there are also cases in which water scarcity results from factors other than a low degree of rule conformance or poor maintenance. Water scarcity can be caused by various environmental and organizational problems that are beyond the immediate control of irrigators. In these cases, the level of water supply in an appropriation resource can be viewed as a contextual variable that affects irrigators' incentives to follow water allocation and maintenance schedules. The chance of having a high level of rule conformance and good maintenance is higher in systems with adequate supplies of water than in those with inadequate supplies of water. Extra coordination costs caused by water scarcity appear to be a major factor affecting collective action in some irrigation systems.

An inadequate supply of water, low degree of rule conformance, and poor maintenance affect the distribution of benefits among irrigators in some systems. In these systems, irrigators who cultivate only land at the tail portion of an appropriation resource usually get a disproportionately smaller amount of water than the headenders. Active cooperation among all irrigators is important for ensuring that no irrigator is consistently disadvantaged.

In the sample of cases, high income variance and the presence of substantial social cleavages among appropriators appear to be conducive to low degrees of rule conformance or poor maintenance. This result, however, has to be treated cautiously because (1) a rather small percentage of the cases report the presence of these two attributes and (2) all the cases characterized by this combination of attributes and outcomes happen to be bureaucratic cases that are also characterized by other features unfavorable for collective action. These two attributes are probably factors that, among others, impose substantial constraints on irrigators' attempts to organize collective action.

Other physical and community attributes including the farmers' degree of dependence on an appropriation resource, the size of the irrigated area, and the number of irrigators may affect the structures of incentives irrigators face and the kinds of institutional arrangements needed to coordinate irrigators' activities. Within the sample of cases, these attributes fail to show any effect on the level of rule conformance and maintenance in an appropriation resource. Individual cases, however, have documented how some of these attributes combine with other factors to affect the level of

TABLE 4.9   Outcomes and Physical and Community Attributes in Community Irrigation Systems

NAME	ADEQ	RULE	MAIN	DISA	INCO	ALTE	CLEA	VARI	SIZE	NUMB
Maruro	no	no	poor	yes	h./l.[a]	no	no	moderate	15	26
Chhahare Khola	no	no	poor	—	—	no	—	—	20	250
Naya Dhara	no	no	poor	—	—	yes	no	—	55	400
Char Hazar	no	no	poor	no	most	yes	no	—	200	—
San Antonio 1	no	no	poor	yes	h./l.	yes	no	—	23	16
San Antonio 2	no	no	poor	yes	h./l.	yes	no	—	7	5
Oaig-Daya	no	no	good	no	h./l.	no	no	low	100	86
Silag-Butir	no	yes	good	yes	h./l.	yes	no	—	114	35
Tanowong T	no	yes	good	no	h./l.	no	no	—	—	200
Sabangan Bato	no	yes	good	no	h./l.	yes	no	—	94	97
Nayband	no	yes	good	no	h./l.	no	no	low	—	40
Calaoaan	yes	yes	good	—	h./l.	yes	no	—	150	71
Felderin	yes	yes	good	yes	h./l.	—	no	moderate	100/—	75
Kheri	yes	yes	good	yes	most	no	no	moderate	260	130
Raj Kulo	yes	yes	good	yes	most	—	no	low	94	159
Saebah	yes	yes	good	—	—	—	no	—	100	90
Chiangmai	yes	yes	good	no	most	yes	yes	high	—/—	167
Zanjera Danum Sitio	yes	yes	good	no	most	no	no	moderate	45/150	23
Tanowong B	yes	yes	good	no	h./l.	no	no	—	—	200

76

	ADEQ	RULE	MAIN	DISA	INCO	ALTE	CLEA	VARI	SIZE	NUMB
Pinagbayanan	yes	yes	good	no	h./l.	yes	no	moderate	20	17
Thulo Kulo	yes	yes	good	no	most	—	no	low	39	105
Takkapala	yes	yes	good	—	—	—	no	—	95	125
Deh Salm	yes	yes	good	—	most	no	yes	moderate	300	80
Bondar Parhudagar	yes	yes	good	no	—	yes	no	low	4	—
Nabagram	yes	yes	good	no	h./l.	no	no	moderate	29	61
Na Pae	yes	yes	good	no	most	no	no	low	64	80
Agcuyo	yes	yes	good	no	most	no	no	—	9	50
Cadchog	yes	yes	good	no	most	no	no	—	3	200
Silean Banua	yes	yes	good	no	most	yes	no	—	120	206

ADEQ = Adequacy (adequacy of water supply)
RULE = Rule conformance (most appropriators follow rules)
MAIN = Maintenance (maintenance of appropriation resource)
DISA = Disadvantaged (any appropriators being consistently disadvantaged)
INCO = Family income (family income derived directly from cultivating crops irrigated by appropriation resource)
ALTE = Alternatives (access to alternative sources of water for irrigation)
CLEA = Cleavages (cultural or social differences among appropriators)
VARI = Income variance (variance in the average annual family income among appropriators)
SIZE = Appropriation size/system size (size of appropriation resource in ha/size of irrigation system in ha)
NUMB = Number (number of appropriators utilizing appropriation resource)

— = Missing in case.
a. Half or less of family income.

cooperation among irrigators. The lack of strong associations between these attributes and certain outcomes in irrigation systems
does not necessarily mean that these attributes are irrelevant for
collective action in irrigation systems. They may produce opposite
effects, depending on the configuration of other contextual factors.

Tables 4.9, 4.10, and 4.11 show the configurations of outcomes
and physical and community attributes of all the forty-seven cases
discussed in this chapter. One may observe from these tables that
a wide diversity of configurations of physical and community attributes are represented in the sample of cases. Cases sharing similar physical and community attributes may be characterized by very
different outcomes. For instance, Mauraro has a pattern identical
to that of Nabagram in terms of their source of family income,
lack of alternative water sources, absence of social cleavages, and
moderate income variance (see Table 4.9). The outcomes of the
two cases, however, are very different.[3]

In conclusion, physical and community attributes create the setting in which irrigators make choices and take actions in an attempt
to improve their welfare. Individuals are capable of shaping outcomes in an irrigation system by constituting their own terms of
cooperation that take into account the constraints and opportunities created by these attributes. The experiences of some of the
cases, such as the one that requires farmers to cultivate fields in
the head, middle, and tail portions of the major canals, demonstrate how institutional arrangements may be established to counter
potential perverse incentives created by some of these physical and
community attributes.

TABLE 4.10 Outcomes and Physical and Community Attributes in Bureaucratic Irrigation Systems

NAME	ADEQ	RULE	MAIN	DISA	INCO	ALTE	CLEA	VARI	SIZE	NUMB
Gondalpur Watercourse	no	no	poor	yes	most	yes	yes	high	200/628,000	95
Area One Watercourse	no	no	poor	yes	most	no	–	high	50/628,000	50
Area Two Watercourse	no	no	poor	yes	most	no	yes	high	33/229,000	10
Dakh Branch Watercourse	no	no	poor	yes	most	yes	yes	high	152/–	56
Dhabi Minor Watercourse	no	no	poor	yes	–	no	yes	high	21/–	60
Punjab Watercourse	no	no	poor	yes	most	yes	yes	moderate	96/–	41
Amphoe Choke Chai	no	no	poor	no	h./l.[a]	yes	no	–	125/12,000	85
Area Three Watercourse	no	yes	poor	yes	most	no	no	moderate	115/33,000	460
Kottapalle	no	yes	good	no	most	no	no	moderate	500/–	800
Nam Tan Watercourse	no	yes	good	no	most	–	no	low	100/2,046	40
Sananeri	no	yes	good	–	most	yes	no	moderate	173/1,172	150
Area Four Watercourse	no	yes	good	no	h./l.	yes	no	low	150/67,670	300
Kaset Samakee	yes	yes	good	no	most	no	no	–	28/12,000	34
El Mujarilin	yes	yes	good	no	h./l.	no	no	moderate	307/208,820	38

ADEQ = Adequacy (adequacy of water supply)
RULE = Rule conformance (most appropriators follow rules)
MAIN = Maintenance (maintenance of appropriation resource)
DISA = Disadvantaged (any appropriators being consistently disadvantaged)
INCO = Family income (family income derived directly from cultivating crops irrigated by appropriation resource)
ALTE = Alternatives (access to alternative sources of water for irrigation)
CLEA = Cleavages (cultural or social differences among appropriators)
VARI = Income variance (variance in the average family income among appropriators)
SIZE = Appropriation size/system size (size of appropriation resource in ha/size of irrigation system in ha)
NUMB = Number (number of appropriators utilizing appropriation resource)

— = Missing in case.
a. Half or less of family income.

79

TABLE 4.11　　　Outcomes and Physical and Community Attributes in Other Irrigation Systems

NAME	ADEQ	RULE	MAIN	DISA	INCO	ALTE	CLEA	VARI	SIZE	NUMB
Hanan Sayoc	no	no	good	—	h./l.[a]	no	no	—	—	600
Lurin Sayoc 1	no	no	good	—	h./l.	no	no	—	—	400
Lurin Sayoc 2	no	no	good	—	h./l.	no	no	—	—	400
Diaz Ordaz Tramo	yes	yes	good	no	most	no	—	low	2/150	—

ADEQ　=　Adequacy (adequacy of water supply)
RULE　=　Rule conformance (most appropriators follow rules)
MAIN　=　Maintenance (maintenance of appropriation resource)
DISA　=　Disadvantaged (any appropriators being consistently disadvantaged)
INCO　=　Family income (family income derived directly from cultivating crops irrigated by appropriation resource)
ALTE　=　Alternatives (access to alternative sources of water for irrigation)
CLEA　=　Cleavages (cultural or social differences among appropriators)
VARI　=　Income variance (variance in the average annual family income among appropriators)
SIZE　=　Appropriation size/system size (size of appropriation resource in ha/size of irrigation system in ha)
NUMB　=　Number (number of appropriators utilizing appropriation resource)

— = Missing in case.
a. Half or less of family income.

80

CHAPTER FIVE

Institutional Arrangements
and Collective Action

Although physical and community attributes affect collective action in an irrigation system, they seldom dictate its success or failure. The long-term viability of an irrigation system depends on rules that can accommodate bounded rationality and safeguard against opportunistic behavior. Cultivators in some irrigation systems are able to shape the structure of their situation by constituting rules that take into account the constraints and opportunities created by various physical and community attributes.

Diverse types of institutional arrangements have been documented in the cases. Although the case authors use different terminology to describe institutional arrangements, it is possible to translate the information they provide into a form that permits the comparison of the structures of institutions across cases. As discussed in Chapter 2, institutional arrangements can be conceptualized as rules that are distinguishable at least at two levels: operational and collective-choice.[1] Operational rules stipulate who can participate as appropriators and providers; what the participants may, must, or must not do; and how they will be rewarded and punished. A second set of rules—collective-choice rules—stipulates the conditions for adopting, enforcing, and modifying operational rules. The distinction between these two levels of rules can serve as a starting point for deciphering information about institutional arrangements discussed in the case studies. By examining rule configurations at each level, one can identify essential differences and similarities underlying various action situations in irrigation systems.

A systematic examination of information contained in the sample of cases reveals the richness and diversity of rule configurations existing in the real world. In this chapter, I discuss some common operational and collective-choice rules found in the cases. I also examine how these rules affect outcomes in irrigation systems under various circumstances and some of the factors that lead to the emergence or adoption of these rules.

Operational Rules

Depending on their physical and community attributes, various irrigation systems pose different types of problems for cultivators. For instance, cultivators in irrigation systems with inadequate supplies of water and poor construction face serious collective-action problems in water allocation and maintenance. Four types of operational rules—boundary, allocation, input, and penalty rules—are important means of coordinating irrigators in water allocation and maintenance in these systems.

Boundary Rules. Boundary rules prescribe the requirements individuals have to meet before appropriating water from an appropriation resource. They define the groups of individuals whose actions will affect one another because of their common relationship to an appropriation resource. Four boundary requirements appear most frequently in the case studies:

1. *Land*: ownership or leasing of land within a specified location

2. *Share*: ownership or leasing of shares, transferable independently of land, to a certain proportion of the water flow or water delivery facilities

3. *Membership*: membership in an organization

4. *Fee*: payment of a certain entry fee each time before withdrawing water

Except for a few cases that do not contain enough information to determine, all cases in the sample are characterized by some forms of boundary requirements. As shown in Table 5.1, uniformity exists among bureaucratic irrigation systems and systems governed by local governments: They all adopt land as the sole boundary requirement.[2] There is, however, a great diversity of boundary requirements among community irrigation systems. In this study, I focus on two boundary requirements—land and shares.

Land and other boundary requirements. A boundary rule will facilitate cooperation among irrigators if it can limit the number of appropriators to such a level that the demand for water does not far exceed the supply. This is because, as discussed in Chapter 4, water scarcity is a major source of conflict in many irrigation systems. Collective-action problems may be aggravated if more cultivators or fields are entitled to receive water than the appropriation resource can support. Many irrigation systems that use land as the sole boundary requirement apparently fail to keep the number of irrigators within limits. As shown in Tables 5.2a and 5.2b, cases that use land as the sole boundary requirement are characterized by a higher incidence of inadequate water supply, lower rule conformance, and poorer maintenance than those that adopt other combinations and types of boundary requirements.

In many of the cases that use land as the sole boundary requirement, water is supposed to be available to all plots within a defined command area. Although this boundary requirement makes water available to more individuals, there are often more irrigators than the source of water can support. A formal policy goal of many bureaucratic irrigation systems in south Asia is to deliver water to as many farmers and as much land as possible. The official command areas in many of these irrigation systems, however, are much larger than can be supported by the sources of water (Palanisami 1982; Repetto 1986). Irrigators in these systems face a high degree of water scarcity and various water allocation and maintenance problems.

In some bureaucratic cases, the government agencies involved may not even have accurate information about their own systems. As documented by Wade (1984), government officials in India have incentives to misrepresent data about their irrigation systems. Officials in the Revenue Department like to report a smaller irrigated area in order to justify collecting a smaller amount of revenue from farmers. Those in the Irrigation Department like to report a larger area in order to claim extra credit for their work. If policy makers cannot have accurate data about the actual volume of water flow in

TABLE 5.1   Boundary Rules Employed by Each Irrigation System

Requirements	Community systems	Bureaucratic systems	Other systems
Shares to resource or flow	Deh Salm Felderin Nayband Thulo Kulo		Diaz Ordaz Tramo Hanan Sayoc* Lurin Sayoc 1* Lurin Sayoc 2*
Shares + membership	Pinagbayanan		
Land	Char Hazar* Chhahare Khola* Chiangmai Mauraro* Naya Dhara* San Antonio 1* San Antonio 2* Tanowong T Tanowong B	Amphoe Choke Chai* Area One Watercourse* Area Two Watercourse* Area Three Watercourse* Area Four Watercourse Dakh Branch Watercourse* Dhabi Minor Watercourse* El Mujarilin Gondalpur Watercourse* Kaset Samakee Kottapalle Nam Tan Watercourse Punjab Watercourse* Sananeri	

Land + other requirements (e.g., membership, fees, shares)	Calaoaan
	Na Pae
	Oaig-Daya*
	Sabangan Bato
	Zanjera Danum Sitio
Different requirements applied to different subgroups	Bondar Parhudagar
	Kheri
	Nabagram
	Raj Kulo
	Silag-Butir
	Silean Banua

* Denotes a case that is negative in either rule conformance or maintenance or both. Cases without asterisks are positive in both rule conformance and maintenance.

85

TABLE 5.2a   Adequacy of Water Supply by
Boundary Rules

	Land as the sole boundary requirement	Other types or combinations of boundary requirements	Total
Adequate water supply	19% 5	75% 12	17
Inadequate water supply	81% 22	25% 4	26
Total	100% 27	100% 16	43

Percentage difference = 56%
Chi-square with continuity correction factor = 11.0
D.F. = 1 $P < 0.001$

an irrigation system or about the actual amount of land it irrigates, they can hardly be expected to deliver sufficient water to cultivators in a timely fashion.

Transferable shares.   A theoretically interesting boundary requirement is the ownership or leasing of shares that can be transferred independently of land. The system of independently transferable water shares tends to encourage efficient uses of water. Martin and Yoder (1986), for example, compare two community irrigation systems—Thulo Kulo and Raj Kulo—in Nepal. Irrigators in Thulo Kulo have a transferable water share system. Farmers who want water are free to purchase water shares from other farmers. Water is therefore likely to go to those who value it the most. In Raj Kulo, water rights during the monsoon rice season are restricted to individuals who cultivate land in a certain part of the village. Even though the supply of water has increased considerably in the past decade, other farmers in need of water cannot benefit because water rights are tied to particular plots within the original command area and are not independently transferable. As a result of this in-

| TABLE 5.2b | Rule Conformance and Maintenance by Boundary Rules | | |

	Land as the sole boundary requirement	Other types or combinations of boundary requirements	Total
Cases rated positive in both rule conformance and maintenance	37% 10	94% 15	25
Cases rated negative in either rule conformance or maintenance or both	63% 17	6% 1	18
Total	100% 27	100% 16	43

Percentage difference = 57%
Chi-square with continuity correction factor = 11.0
D.F. = 1   P < 0.001

flexible boundary requirement, excess water is diverted to a drain rather than used to cultivate crops outside the original command area.

In spite of the efficient feature of transferable water rights, only a few cases in the sample report the use of transferable water rights. In four cases, transferable shares are the sole boundary requirement; in another case, they are used in conjunction with membership (see Table 5.1). There are two possible reasons for the rarity of this form of institutional arrangement. One possibility is that transferable water rights are feasible only under very special circumstances. Glick (1970), for example, argues that more technological control is usually needed to enforce the transferable rights system. The cases in the sample, however, do not appear to support this argument: The share arrangements in Deh Salm, Nayband, and Felderin work effectively with little technological and organizational control. In all three systems, each water share corresponds to a fixed time slot in a distribution cycle. Because all share owners know their time slots, they are supposed to divert water to their fields from certain outlets during those periods.

In Thulo Kulo, a somewhat more sophisticated system is at work. The water rights arrangement is accomplished through the use of *saachos*—beams with several notches of equal depth but various widths cut into the top. A *saacho* is "installed in a canal such that all the water flows through the notches causing the flow to be divided proportionally relative to the ratio of the widths of the notches" (Martin and Yoder 1983a, 14). Adjusting the size of the notch lets water be distributed to individual farmers according to the amounts of water rights they hold. Although such an arrangement is more sophisticated than the arrangements in Deh Salm, Nayband, and Felderin, it requires only very simple construction and operating procedures.

The second possible reason only a few cases report the presence of transferable water rights is that some case authors may have failed to recognize the property rights arrangements in some of the irrigation systems. Coward writes:

> The simple technology of traditional irrigation works and the apparent casualness with which they operate often mislead outsiders into assuming that little of value exists. The untrained observer can easily fail to extract from the rude weirs and rough canal structures the sometimes intricate property relations which such prior investments have created. (1986, 226)

The identification of property rights in irrigation systems is difficult not only for "the untrained observer" but also for experienced researchers. Robert Yoder, for example, indicated that he was not aware of the water share system in Thulo Kulo until after he had spent six months in the village.[3] Unless careful observation has been made, the detailed property rights arrangements in an irrigation system may not be readily apparent.

Allocation Rules. Whereas boundary rules prescribe the requirements one must fulfill before taking water from an appropriation resource, allocation rules stipulate the procedures and bases by which individuals can withdraw water from an appropriation resource. Allocation rules determine how much water one can get and when one can get it. A wide diversity of water allocation rules can be found in the cases. Three types of procedures—fixed percentage, fixed time slot, and fixed order—are frequently used in water allocation. Each may be based on different premises, such as amount of land held, amount of water needed to cultivate existing crops, number of shares held, location of field, or official discretion. An allocation rule, for instance, may require each irrigator to appropriate water in specific time slots. The length of the slot assigned to each irrigator may be determined by the number of water shares held; for example, the greater the number of shares one holds, the longer the time slot to which one is entitled.

All of the cases in the sample, except for three, have some form of allocation rules. Although the presence of allocation rules does not guarantee success, the three cases that lack allocation rules—Mauraro, Chhahare Khola, and San Antonio 2—face various kinds of water allocation and maintenance problems. All three cases have an inadequate supply of water, and conflicts arise frequently among appropriators in the absence of any allocation rules.[4]

Fixed time slots and other allocation procedures. In an irrigation system, more than one set of allocation rules may be used for different occasions. In many systems, a more restrictive set of allocation rules is used during certain periods in a year and a less restrictive set is used during other periods. Table 5.3 shows the most restrictive sets of rules used in the sample of cases. Among the three types of procedures, fixed time slots are the most commonly used. Assigning irrigators fixed time slots may be an economical way of distributing water. As discussed earlier, this method of distribution is successful in Deh Salm, Nayband, and Felderin. In these systems, all irrigators know their time slots, and each irrigator will show up and divert water to his own plots from certain outlets when his time slot begins.

This water distribution procedure, however, has a potential problem: If the water flow is erratic, an irrigator owning a share for a particular time slot is still uncertain about his supply of water. Dhabi Minor Watercourse, for example, is located in a bureaucratic irrigation system where irrigators are assigned time slots in different water distribution cycles within a watercourse. At the system level, water supplies to various watercourses are determined by yet another water distribution cycle. Because of a lack of coordination between distribution cycles at the two levels, an irrigator

TABLE 5.3   The Most Restrictive Allocation Rule Employed by Each Irrigation System

	Basis of water distribution				
	Land/needs	Shares	Location	Official discretion	Other
Fixed percentage	"Raj Kulo"	"Na Pae" "Thulo Kulo" "Zanjera Danum Sitio"		San Antonio 1*	
Fixed time slots	Oaig-Daya* Naya Dhara* [Area One Watercourse]* [Area Three Watercourse]* [Area Four Watercourse] [Dhabi Minor Watercourse]*	Deh Salm Felderin Nayband	Calaoaan	Chiangmai Kheri "Na Pae" Pinagbayanan Hanan Sayoc* Lurin Sayoc 2* [Amphoe Choke Chai]* [Area Two Watercourse]* [Area Three Watercourse]* [Kaset Samakee] [Nam Tan Watercourse]	Lurin Sayoc 1* [Punjab Watercourse]* [Gondalpur Watercourse]*

90

Fixed order	Sabangan Bato	Cadchog Char Hazar Silag-Butir "Zanjera Danum Sitio"	Tanowong T Tanowong B "Thulo Kulo" "Raj Kulo"	Agcuyo Nabagram
No rule	[Kottapalle] [Sananeri]			Chhahare Khola* Mauraro* San Antonio 2*

NOTE: Bureaucratic systems are shown in brackets. Cases that use more than one allocation rule are shown in quotation marks.
* Denotes a case that is negative in either rule conformance or maintenance or both. Cases without asterisks are positive in both rule conformance and maintenance.

TABLE 5.4   Rule Conformance and Maintenance by Allocation Rules

	Fixed time slots as the sole distribution procedure	Other types or combinations of distribution procedures	Total
Cases rated positive in both rule conformance and maintenance	43% 10	93% 14	24
Cases rated negative in either rule conformance or maintenance or both	57% 13	7% 1	14
Total	100% 23	100% 15	38

Percentage difference = 50%
Chi-square with continuity correction factor = 7.7
D.F. = 1   P < 0.01

assigned a particular time slot may fail to get any water if no water is scheduled to flow into the watercourse during that time. Irrigators in Dhabi Minor Watercourse therefore face a high degree of uncertainty about their water supplies, which in turn affects their willingness to cooperate in water allocation and maintenance.

Cases using fixed time slots as the sole distribution procedure are characterized by a higher incidence of problems in rule conformance or maintenance than those using other types or combinations of distribution procedures (see Table 5.4). Distributing water by fixed time slots may require less administrative costs than other distribution procedures. Serious collective-action problems may, however, arise if the procedure is used without considering whether it is compatible with other institutional and physical attributes of the appropriation resource. The example of Dhabi Minor Watercourse is a case in point. Within the sample, this kind of incompatibility appears to arise mostly in bureaucratic irrigation systems: Of the twelve cases that use fixed time slots as the sole distribution procedure and have problems in rule conformance or maintenance, eight are bureaucratic.

Adjusting allocation rules to changes in supply. In some irrigation systems, demands for water may temporarily exceed supplies during dry seasons or some growth stages of the crops. Water allocation rules in these systems may have to be adjusted in response to changes in the balance between supply and demand. Within the sample, nineteen cases are reported to have two sets of allocation rules. All except one have more restrictive rules during times of scarcity than during times of abundance. In some, appropriators are permitted to withdraw water freely during periods of abundance; some types of turns or time schedules are used when water gets scarce. In Cadchog in the Philippines, for example, water flows from the canal to plots through various take-off points. During the wet season, all the take-off points are kept open most of the time. Water is allowed to reach as many plots as the available water flow will serve. During the dry season, a time of scarcity, water is distributed in a certain order. The entire irrigated area is divided into four sections, each of which takes turns in obtaining water.

In some other cases, officials or monitors begin to exercise discretion in setting up time schedules or turns for water allocation when the supply of water decreases. In Tanowong of the Philippines, for example, irrigators are allowed to withdraw water freely from the system during the rainy season when water is abundant. During the dry season from February to April, eight to twelve water distributors are selected by appropriators to "take over the task of systematically distributing the water as fairly as they can to the different fields" (Bacdayan 1980, 176). The involvement of officials in water allocation is a way to reduce conflicts or chances of rule violations among irrigators. Provided that water distributors are held accountable to irrigators, irrigators can be relieved of the trouble of having to spend time and energy guarding their own water allotments against theft.

In Sananeri Tank in India, allocation rules are relaxed at times of extreme water scarcity. Sananeri Tank is located in a large bureaucratic irrigation system. A water users' association exists to govern water appropriation activities from the tank. During most of the year, appropriators are allowed to withdraw water from the tank freely. In the dry season, six "water spreaders" appointed by the association take over the water allocation job. These water spreaders, however, stop distributing water whenever the water level in the tank is too low to irrigate the entire appropriation area. They will notify the cultivators of the fact. After that, cultivators may take any available water for their own use. In this situation, cultivators with fields near the tank have an advantage in obtaining water over those with fields farther away. Meinzen-Dick argues that this arrangement reflects the principle that those who receive water should pay for the costs of its acquisition:

> If tank water cannot be used to serve the entire ayacut, it would be unfair to those who did not receive water if common ayacut resources were used to distribute water to the headenders. The switch from collective to individual distribution ensures that all cultivators in the ayacut receive roughly equal benefit from the water distribution activities of the organization. This also lifts the burden of expense and effort for applying water to the fields from the association (all cultivators) to those who receive additional water. (1984, 76)

This arrangement has not created too much conflict among irrigators in Sananeri because many irrigators have access to an alternative source of water—private wells; those who do not own wells may purchase well water from those with electric pumpsets at an hourly rate.

These cases show that different rules may be adopted to coordinate water allocation under various circumstances. Even holding all other conditions constant and allowing only changes in water supplies, as within one appropriation resource, allocation rules have to be adjusted from time to time to accommodate different degrees of water scarcity.

Input Rules. Input rules stipulate the types and amounts of resources required of each cultivator. There are four major types of inputs an irrigator may be required to contribute: (1) regular water tax; (2) labor for regular maintenance; (3) labor for emergency repair; and (4) labor, money, or materials for major capital investment. As shown in Table 5.5, regular water taxes are required of irrigators in half the community cases, but in almost all bureaucratic cases. The presence or absence of regular water taxes does not appear to have any definite effect on the outcomes of an irrigation system. With a few exceptions, almost all the cases require some labor inputs from irrigators. Direct labor inputs from irrigators may or may not solve maintenance problems in an appropriation resource, depending on whether the labor force is effectively organized and motivated to do the job.

Although capital investments are required in most community cases, they are used in only half the bureaucratic cases. It appears that irrigators in bureaucratic systems are more motivated to cooperate in water allocation and maintenance if they are involved in

TABLE 5.5   Input Rules Employed in Each Irrigation System

	Water tax	Regular labor	Emergency labor	Capital investment
COMMUNITY SYSTEMS				
Mauraro*	no	yes	no	no
Chhahare Khola*	no	yes	yes	yes
Naya Dhara*	no	yes	yes	yes
Char Hazar*	no	yes	yes	no
San Antonio 1*	yes	yes	yes	yes
San Antonio 2*	yes	yes	yes	yes
Oaig-Daya*	no	yes	yes	yes
Silag-Butir	yes	yes	yes	yes
Tanowong T	—	yes	yes	no
Sabangan Bato	no	yes	yes	—
Nayband	—	—	—	—
Calaoaan	no	yes	yes	yes
Felderin	no	yes	yes	yes
Kheri	yes	yes	yes	—
Raj Kulo	yes	yes	yes	yes
Saebah	—	yes	yes	yes
Chiangmai	yes	yes	yes	yes
Zanjera Danum Sitio	—	yes	yes	—
Tanowong B	—	yes	yes	yes
Pinagbayanan	yes	yes	yes	yes
Thulo Kulo	yes	yes	yes	yes
Takkapala	—	yes	yes	yes
Deh Salm	no	no	no	no
Bondar Parhudagar	yes	yes	yes	yes
Nabagram	yes	yes	yes	—

— = Missing in case.
* Denotes a case that is negative in either rule conformance or maintenance or both. Cases without asterisks are positive in both rule conformance and maintenance.

continued on next page

TABLE 5.5 *continued*

	Water tax	Regular labor	Emergency labor	Capital investment
COMMUNITY SYSTEMS				
Na Pae	no	yes	yes	yes
Agcuyo	—	yes	yes	no
Cadchog	no	yes	yes	—
Silean Banua	yes	yes	yes	yes
BUREAUCRATIC SYSTEMS				
Gondalpur Watercourse*	yes	yes	yes	no
Area One Watercourse*	yes	yes	yes	no
Area Two Watercourse*	yes	no	no	no
Dakh Branch Watercourse*	yes	yes	yes	no
Dhabi Minor Watercourse*	yes	no	no	no
Punjab Watercourse*	yes	yes	yes	no
Amphoe Choke Chai*	yes	yes	yes	yes
Area Three Watercourse*	yes	yes	yes	yes
Kottapalle	yes	no	no	yes
Nam Tan Watercourse	yes	—	—	no
Sananeri	yes	yes	yes	yes
Area Four Watercourse	yes	yes	yes	yes
Kaset Samakee	yes	yes	yes	yes
El Mujarilin	—	yes	yes	yes
OTHER SYSTEMS				
Hanan Sayoc*	—	yes	yes	—
Lurin Sayoc 1*	—	yes	yes	—
Lurin Sayoc 2*	—	yes	yes	—
Diaz Ordaz Tramo	yes	yes	yes	no

— = Missing in case.
* Denotes a case that is negative in either rule conformance or maintenance or both. Cases without asterisks are positive in both rule conformance and maintenance.

capital investments in their own appropriation resources. Five of the seven cases that require irrigators to contribute capital investments are characterized by both a high level of rule conformance and good maintenance. Only one of the seven cases without requirements for capital investments is characterized by both a high level of rule conformance and good maintenance.

Bases for Labor Inputs. Two major types of rules for labor inputs can be identified. One type simply requires equal contribution from all the appropriators. The other requires labor inputs from appropriators roughly in proportion to the benefits each obtains from the resource (for example, proportional to one's share of the resource, to the amount of land cultivated, or to the amount of water needed for cultivation).

Complete information about rules for regular labor inputs is available in only seventeen of the cases (see Table 5.6). Cases using proportional rules are characterized by a higher incidence of rule conformance and good maintenance than those using equal rules (see Table 5.7). The relationship, however, is insignificant (with a level of confidence equal to 0.47).

This result does not fully support the argument that labor obligations ought to be proportional to expected benefits in order to be effective (see Chapter 2). Although the insignificant result may stem from the limited number of cases available, an additional

---

TABLE 5.6   Bases of Regular Labor Input Rules

Proportional basis	Equal basis
Chiangmai	
Thulo Kulo	Cadchog
Raj Kulo	Calaoaan
Zanjera Danum Sitio	Na Pae
Oaig-Daya*	Chhahare Khola*
Mauraro*	Naya Dhara*
Diaz Ordaz Tramo	Hanan Sayoc*
	Lurin Sayoc 1*
[Sananeri]	Lurin Sayoc 2*
[Gondalpur Watercourse]*	

NOTE: Bureaucratic systems are shown in brackets.
* Denotes a case that is negative in either rule conformance or maintenance or both. Cases without asterisks are positive in both rule conformance and maintenance.

TABLE 5.7   Rule Conformance and Maintenance by Regular Labor Input Rules

	Proportional rules	Equal rules	Total
Cases rated positive in both rule conformance and maintenance	67% 6	38% 3	9
Cases rated negative in either rule conformance or maintenance or both	33% 3	62% 5	8
Total	100% 9	100% 8	17

Percentage difference = 29%
Chi-square with continuity correction factor = 0.51
D.F. = 1   P > 0.1

factor—maintenance intensity—may make equal rules more effective than proportional rules in some circumstances. Maintenance intensity can be roughly measured by dividing the total number of person-days of labor per year mobilized to maintain the production, distribution, and appropriation resources by the total number of appropriators in an appropriation resource. Only eleven of the cases report information about both maintenance intensity and labor input rules for maintenance (see Table 5.8). For the seven cases that require equal labor contribution, the average maintenance intensity is 2.3 days per person per year. For the four that require proportional labor contribution, the average is 17.7 days per person per year. One possible inference from this limited amount of information is that systems with a higher maintenance intensity tend to adopt the proportional rule for labor inputs, while systems with lower maintenance intensity tend to adopt the equal contribution rule.

Administrative cost appears to be a factor that makes equal contribution rules a better choice than proportional rules in some circumstances. For the proportional rule to be enforced, resources

TABLE 5.8	Regular Labor Input Basis and Maintenance Intensity	
Cases	Basis for regular labor input	Maintenance intensity (person-days of labor per cultivator per year)
Cadchog	Equal	4.0
Na Pae	Equal	3.5
Chhahare Khola*	Equal	2.0
Hanan Sayoc*	Equal	1.5
Lurin Sayoc 1*	Equal	1.5
Lurin Sayoc 2*	Equal	1.5
Naya Dhara*	Equal	2.0
(Average)		(2.3)
Chiangmai	Proportional	28.0
Thulo Kulo	Proportional	16.7
Raj Kulo	Proportional	11.0
Oaig-Daya	Proportional	15.0
(Average)		(17.7)

* Denotes a case that is negative in either rule conformance or maintenance or both. Cases without asterisks are positive in both rule conformance and maintenance.

must be expended in counting, recording, and organizing various contributions from different appropriators. For systems that require only two or three days of work from each appropriator every year, the potential benefits of proportional rules could easily be offset by the cost of implementing them. Systems with higher maintenance intensity, on the other hand, may gain more from the proportional rules than they expend in the administrative costs.

This argument is supported by the emergency labor rules found in the sample of cases (see Table 5.9). Within the sample, cases using proportional rules do not appear to be more likely to have a high degree of rule conformance and maintenance than those using equal rules (see Table 5.10). In eight of the cases, equal contribution rules are used for emergency labor inputs. These resources are all located

TABLE 5.9   Bases of Emergency Labor Input Rules

Proportional basis	Equal basis
Chiangmai	Calaoaan
Zanjera Danum Sitio	Thulo Kulo
Naya Dhara*	Raj Kulo
Oaig-Daya*	Na Pae
	Chhahare Khola*
Diaz Ordaz Tramo	
	Hanan Sayoc*
[Sananeri]	Lurin Sayoc 1*
[Gondalpur Watercourse]*	Lurin Sayoc 2*

NOTE: Bureaucratic systems are shown in brackets.
* Denotes a case that is negative in either rule conformance or maintenance or both. Cases without asterisks are positive in both rule conformance and maintenance.

TABLE 5.10   Rule Conformance and Maintenance by Emergency Labor Input Rules

	Proportional rules	Equal Rules	Total
Cases rated positive in both rule conformance and maintenance	57% 4	50% 4	8
Cases rated negative in either rule conformance or maintenance or both	43% 3	50% 4	7
Total	100% 7	100% 8	15

Percentage difference = 7%
Chi-square with continuity correction factor = 0.06
D.F. = 1   P > 0.5

in steep terrain. The water distribution system can be destroyed easily by sudden increases in water flow in rainy or stormy weather. Speedy repair is needed to ensure the continual functioning of the entire system. Equal contribution rules allow labor to be mobilized rapidly. The prospect of losing the entire irrigation system can be a sufficient incentive for the cultivators to participate in the joint endeavor.

Penalty Rules. Individuals may have little incentive to follow allocation and input rules if rule breakers are not liable to any penalty. Penalty rules can take many forms, including incarceration, loss of appropriation rights, fines, and community shunning. Although no single set of penalty rules can guarantee cooperation, the absence of penalty rules makes cooperation difficult. As shown in Table 5.11, seven cases in the sample use none of these four penalties, and all seven have problems in rule conformance or maintenance.

The effectiveness of a penalty rule in deterring rule violations depends on various contextual attributes of an irrigation system. Incarceration, for example, is not commonly used because it requires the involvement of law enforcement agencies, which are not readily available in most irrigation systems. The use of incarceration as a penalty against rule violators is reportedly used in two of the cases in the sample. In these cases, it is enforced and carried out by government authorities, but it is used mostly as a threat. In Diaz Ordaz in Mexico, for example, a local official called a *sindico* is in charge of water distribution among different irrigation sections. The *sindico* is the local representative of the state government. He shares responsibility with the local judges in adjudicating disputes among irrigators. If the *sindico* considers a dispute especially serious, he may refer the case to the district court where the disputants may face heavy legal expenses and possible incarceration. This threat gives the *sindico* great authority in settling disputes among irrigators.

Temporary loss of appropriation rights is a serious penalty for farmers who rely on irrigated agriculture as a major source of income. It is reportedly used as a penalty in ten cases in the sample. In most cases, it is used only under special circumstances. In Oaig-Daya in the Philippines, a farmer is required to pay a fine for being absent from a general meeting. If the farmer refuses to pay the fine, officials of the farmers' association will compel the farmer to pay the fine by depriving him of water. In Kaset Samakee, Raj Kulo, Hanan Sayoc, Lurin Sayoc 1, and Lurin Sayoc 2, a farmer is denied irrigation water if he refuses to work in the periodic cleaning or pay a fine. In Nabagram in Bangladesh, farmers must pay irrigation fees in advance before getting the initial water for land preparation and transplanting.

TABLE 5.11	Penalty Rules Employed in Irrigation Systems			
	Incarceration	Loss of entry rights	Fines	Community shunning
COMMUNITY SYSTEMS				
Mauraro*	no	no	no	no
Chhahare Khola*	no	no	no	no
Naya Dhara*	no	no	no	no
Char Hazar*	no	no	yes	yes
San Antonio 1*	no	no	no	no
San Antonio 2*	no	no	no	no
Oaig-Daya*	no	yes	yes	yes
Silag-Butir	no	yes	yes	—
Tanowong T	no	no	yes	yes
Sabangan Bato	—	—	yes	—
Nayband	no	no	—	—
Calaoaan	—	no	yes	—
Felderin	no	no	no	—
Kheri	no	yes	yes	no
Raj Kulo	no	yes	yes	yes
Saebah	—	—	—	—
Chiangmai	no	no	yes	yes
Zanjera Danum Sitio	—	—	—	—
Tanowong B	no	no	yes	yes
Pinagbayanan	no	no	yes	yes
Thulo Kulo	no	no	yes	yes
Takkapala	—	—	—	—
Deh Salm	no	no	no	—
Bondar Parhudagar	no	no	—	yes
Nabagram	no	yes	no	yes

— = Missing in case.
* Denotes a case that is negative in either rule conformance or maintenance or both. Cases without asterisks are positive in both rule conformance and maintenance.

continued on next page

TABLE 5.11 *continued*

	Incarceration	Loss of entry rights	Fines	Community shunning
COMMUNITY SYSTEMS				
Na Pae	no	no	yes	yes
Agcuyo	no	no	yes	yes
Cadchog	no	no	no	yes
Silean Banua	no	no	yes	yes
BUREAUCRATIC SYSTEMS				
Gondalpur Watercourse*	no	no	no	no
Area One Watercourse*	no	no	—	—
Area Two Watercourse*	no	no	—	no
Dakh Branch Watercourse*	no	no	yes	no
Dhabi Minor Watercourse*	no	—	—	—
Punjab Watercourse*	—	no	yes	no
Amphoe Choke Chai*	no	no	no	no
Area Three Watercourse*	no	no	—	yes
Kottapalle	no	no	yes	yes
Nam Tan Watercourse	—	—	—	—
Sananeri	no	no	—	—
Area Four Watercourse	no	no	—	yes
Kaset Samakee	no	yes	no	no
El Mujarilin	yes	yes	—	yes
OTHER SYSTEMS				
Hanan Sayoc*	no	yes	yes	no
Lurin Sayoc 1*	no	yes	yes	no
Lurin Sayoc 2*	no	yes	yes	no
Diaz Ordaz Tramo	yes	no	—	—

— = Missing in case.
* Denotes a case that is negative in either rule conformance or maintenance or both. Cases without asterisks are positive in both rule conformance and maintenance.

Fines are the most commonly used penalty within the sample of cases: They are reportedly used in twenty-one cases. In many of these cases, fines are a routine way to substitute for direct labor inputs. Although this kind of arrangement helps to ensure that no one can free ride on others' contributions, it also allows those who have other obligations or dislike manual labor to bail out without undermining the cooperative arrangement among irrigators. In Thulo Kulo, for example, cash fines are levied against those who are absent from maintenance work. The fine for missing a day of ordinary maintenance is equivalent to the wage for a day of work. The fine for missing emergency maintenance is set higher to ensure a higher rate of attendance. The fine is, however, reduced if the member has some legitimate reasons like illness or absence from the village when the emergency is declared.

Community shunning is a more subtle way of punishing rule breakers. It is reportedly used as a penalty against rule breakers in eighteen cases in the sample. Community shunning can be an effective penalty if appropriators have a high level of consensus about the legitimacy and the importance of the operational rules in use. If such a consensus is established, community shunning alone can be an effective check on free riders. In Cadchog, for example, which is characterized by a high degree of rule conformance and adequate maintenance, community shunning is the only form of penalty that can be imposed on rule violators.

Collective-Choice Arrangements

Operational rules are neither self-generating nor self-enforcing. They depend on individuals who, in coordination with one another, formulate and enforce them. Small numbers of individuals might be able to formulate and enforce rules without any explicit collective-choice arrangements. In most cases, however, such arrangements are needed to formulate, modify, and enforce operational rules. Whether participants in an irrigation system can develop and sustain an effective set of operational rules often depends on the kind of collective-choice arrangements available. If properly constituted, these arrangements enable participants to respond to changes by facilitating rule adoption and alteration. They also help settle disputes and sustain mutually productive relationships among participants by monitoring and sanctioning rule violations and official abuses.

Explicit collective-choice arrangements are absent in seven cases of the sample and present in forty. In this section, I first discuss the physical and community attributes of the cases without explicit collective-choice arrangements and the manner in which these attributes affect their outcomes. I then discuss the cases with explicit collective-choice arrangements and the manner in which different types of collective-choice arrangements affect their outcomes.

Cases without Explicit Collective-Choice Arrangements. Although collective-choice arrangements are usually needed to coordinate rule formulation and enforcement, irrigators in small-scale systems may be able to develop and sustain operational rules without recourse to explicit collective-choice arrangements. Explicit collective-choice arrangements are absent in seven of the community cases in the sample. Four of the cases are characterized by a high degree of rule conformance and good maintenance. None of them need extensive maintenance; irrigators in these systems are able to arrange to maintain their systems without specific leadership. The other three are characterized by a low degree of rule conformance and poor maintenance.

In the four cases characterized by a high degree of rule conformance and good maintenance—Nayband, Felderin, Deh Salm, and Agcuyo—local irrigators have been able to develop and sustain operational rules without explicit collective-choice arrangements. Mechanisms to coordinate their activities have evolved as a result of their mutual adjustment rather than conscious design. Felderin in the Swiss Alps provides an example in which rules have been evolved and sustained through time with neither affirmation nor enforcement by any explicit collective-choice arrangement. Netting describes the evolution of the water share system in Felderin as follows:

> Little organized community activity was required to build the main channels, and as individuals extended the ditches into new meadow areas, they worked out limited and idiosyncratic agreements for water sharing. Though water rights accompany land, they are seldom specified in the elaborate deeds of land transfer that appear from the seventeenth century onward. It is reasonable to assume that inheritance and sale have subdivided properties and at times led to exchange and limited reapportionment of water rights. (1974, 73)

A similar process appears to have developed in Nayband, Deh Salm, and Agcuyo, all of which have effective operational rules governing water allocation and maintenance.

These four cases share certain attributes. First, they involve either a small number of appropriators or a small irrigated area:

Nayband has a population of around eighty families; Felderin irrigates nineteen hectares of fields; both Agcuyo and Deh Salm serve fifty families. Because of the small number of individuals or the small area involved, irrigators can monitor one another. Second, the water allocation processes in these cases are self-regulating. Three of the cases—Nayband, Deh Salm, and Felderin—adopt water share arrangements that allocate water to shareholders according to prespecified time schedules. Shareholders have incentives to monitor one another's activities in order to protect their own water allotments. The water supply in Agcuyo is abundant during the wet season. During the dry months, farmers have to take turns to get water, but there is always sufficient water for all. Third, the four resources require only minimal maintenance. Nayband and Felderin can remain in good shape as long as appropriators use them with care; little maintenance work is required. In Agcuyo, farmers can coordinate in cleaning without specific leadership. In Deh Salm, the appropriation resource requires only limited maintenance. Even cleaning the main canals can be accomplished by individuals.

Although appropriators in these cases are able to manage the day-to-day operation of the irrigation system without explicit collective-choice arrangements, they may face serious problems when major challenges arise. A case in point is Deh Salm, where water is delivered to the appropriation resource through a *qanat* (tunnel). Once the *qanat* has been built, it can work for decades without maintenance. If the *qanat* collapses, however, substantial resources are needed to repair it. During the 1920s, the *qanat* was destroyed by an unusually heavy rainfall, and people in Deh Salm were unable to organize themselves to repair it. They had to rely on people from outside to do the job for them. In so doing, they had to give up some of their water shares to these outsiders. If no outsider had been interested in investing, these farmers might have lost their entire irrigation system.

Although farmers in some small-scale irrigation systems might be able to govern and sustain their systems without explicit collective-choice arrangements, there is no guarantee of success. In the three other cases that are not governed by explicit collective-choice entities—Mauraro, Chhahare Khola, and Naya Dhara—appropriators fail to coordinate their activities and face serious problems in water allocation and maintenance. Mauraro in the Philippines, for example, serves only fifteen hectares of riceland cultivated by twenty-six farmers in the wet season. No organization exists to govern their water allocation and maintenance activities. In order to get water, each farmer has to create his own take-off points along the canal and guard the setup against tampering by other farmers. Conflicts over water allocation arise frequently. Although farmers are supposed to be responsible for cleaning and repairing parts of the system on or bordering their farms, they have developed no rules regarding maintenance of the entire system. The system as a whole is poorly maintained. This case shows that even a small number of irrigators in a small irrigation system may fail to develop effective operational rules without recourse to explicit collective-choice arrangements.

Cases with Explicit Collective-Choice Arrangements. Forty of the cases in the sample have explicit collective-choice arrangements that set the terms and conditions for the formulation, enforcement, and alteration of operational rules. Twenty-five of these cases are characterized by a high degree of rule conformance and good maintenance. The other fifteen have problems in rule conformance or maintenance.

As discussed in Chapter 2, the effectiveness of a set of collective-choice arrangements depends on its ability to help formulate rules that meet the needs of appropriators, detect and sanction rule violations, and hold officials accountable to irrigators. These functions can be discharged by several collective-choice arrangements. First, the direct involvement of irrigators in major collective decisions is important to ensure that the decisions reflect their interests and needs. A way to achieve this is to allow all irrigators in an appropriation resource to participate in major decisions concerning the resource.

Second, individuals have little incentive to comply with a set of rules unless they believe their noncompliance will result in substantial punishment. For operational rules to be enforced, it is necessary to develop mechanisms that can detect and provide sanctions against noncompliance. Mutual monitoring may be effective in a small group of farmers, but in a larger group the provision of monitoring is itself subject to the free-riding problem. Specialized officials or monitors have to be appointed to enforce rules.

Third, officials vested with special prerogatives in rule formulation and enforcement are in a position to abuse their powers by interpreting rules to their own advantage or demanding favors from individual irrigators. Opportunistic behavior by officials is a danger in any collective-choice entity. In order to ensure accountability, rules are needed to stipulate how irrigation officials are to be selected and removed, to whom they have to report, and how they are to be compensated for their services. Officials are likely to be responsive to the needs of irrigators if (1) their tenures are subject to periodic votes by the irrigators, (2) they have to report to the irrigators periodically, and (3) their salaries depend on direct contributions from the irrigators.

Within the sample of cases, collective-choice entities governing most of the community cases are characterized by the above-mentioned collective-choice arrangements. Government agencies in bureaucratic cases, on the other hand, are mostly characterized by collective-choice arrangements that are unfavorable to effective rule formulation, rule enforcement, and official accountability. Among cases with explicit collective-choice arrangements, community cases have a higher incidence of rule conformance and good maintenance than bureaucratic cases (see Table 5.12). Differences in collective-choice arrangements help to explain why the bureaucratic cases are more likely to have inferior outcomes than the community cases in the sample.

Community systems. Collective-choice entities are present in twenty-one of the community cases in the sample. Ten of these cases are governed by irrigators' associations that are responsible only for activities related to the irrigation systems. In Bondar Parhudagar, for example, the irrigation organization's activities are confined to irrigation matters, including the collection of irrigation levies and the maintenance of irrigation canals. In nine other cases, some villagewide or communal organizations that have other responsibilities besides irrigation are responsible for governing the irrigation systems. For instance, the council of elders in Kheri is responsible for governing irrigation matters in the village. It may also constitute itself as a court of law and in formal meetings function as the executive authority for the entire village.

In another case, Na Pae in Thailand, two collective-choice entities are involved in governing one appropriation resource. The irrigation organization, headed by two leaders, is responsible for calling meetings, supervising the operation and maintenance of the

TABLE 5.12   Rule Conformance and Maintenance by
Organizational Form in Irrigation Systems with
Collective-Choice Arrangements

	Community system	Bureaucratic system	Total
Cases rated positive in both rule conformance and maintenance	82% 18	43% 6	24
Cases rated negative in either rule conformance or maintenance or both	18% 4	57% 8	12
Total	100% 22	100% 14	36

Percentage difference = 39%
Chi-square with continuity correction factor = 4.2
D.F. = 1   P < 0.05

irrigation facilities, and representing the organization to government agencies. Because all members of the irrigation organization are from a nearby village, the leader of that village can intervene both directly and indirectly in irrigation matters in Na Pae. The village leader is responsible for supervising the elections of leaders of the irrigation organization, which are held every four years in the village temple. He also helps to obtain assistance from government agencies and mobilize cash and labor for special repair and construction projects in the irrigation system.

Nineteen of the twenty-one community cases are simple systems. Each of these nineteen cases, except Na Pae mentioned above, is governed by one collective-choice entity. The other two cases, Zanjera Danum Sitio and Chiangmai, are complex irrigation systems in which multiple levels of collective-choice entities are involved in governing the systems. Zanjera Danum Sitio, for example, is an appropriation resource within a larger irrigation system. Within each *sitio* (basic appropriation area), an irrigation leadership is selected by its own shareholders. The leadership is responsible for coordinating irrigation activities within the *sitio*. Above the sitio level, the irrigation system is divided into three major branches, each governed by a branch headman who is selected by the total membership of the entire system and coordinates irrigation activities within the branch. At the system level is an organization that deals with important external actors and coordinates activities within the system at both the branch and *sitio* levels. The leaders of this system-level organization are selected annually by the total membership of the entire irrigation system. These system-level leaders, along with those at the branch and *sitio* levels, "are able to act in a manner that considers both the location-specific conditions of an individual *sitio* and the corporate needs of several, or all, *sitios*" (Coward 1979, 32).

Five questions about collective-choice rules in these twenty-one community cases were asked in the coding form:

- *General meetings*: Can appropriators participate in general meetings to express their needs and concerns to those officials of this organization who make collective-choice decisions in relation to the resource?

- *Monitors*: Are specialized officials or monitors appointed by appropriators to enforce operational rules?

- *Executive election*: Are the chief executives selected through direct or indirect elections by appropriators?

- *Executive tenure*: For how long can an individual serve as a chief executive?

- *Executive pay*: Are the chief executives paid?

In most of the community cases, major collective decisions concerning an appropriation resource are made in general meetings that involve most irrigators using the resource (see Table 5.13). In Thulo Kulo and Raj Kulo, for example, general meetings for the entire membership of the irrigators' organizations are held in mid-May. At the meetings, plans for major annual maintenance are drawn, new officials are elected if necessary, and operational rules for the coming season are reviewed and amended if needed. In Raj Kulo,

TABLE 5.13   Collective-Choice Rules in Collective-Choice Entities: Community Irrigation Systems

	General meetings held	Monitors to enforce rules	Elected executives	Tenure of executives	Paid executives
Char Hazar*	yes	yes	yes	fixed[a]	yes
San Antonio 1*	yes	yes	yes	fixed	no
San Antonio 2*	yes	no	yes	fixed	no
Oaig-Daya*	yes	yes	yes	var.[b]	no
Silag-Butir	yes	yes	yes	var.	yes
Tanowong T	yes	yes	yes	—	yes
Sabangan Bato	yes	no	yes	var.	yes
Calaoaan	yes	yes	yes	var.	no
Kheri	—	yes	no	life[c]	yes
Raj Kulo	yes	yes	yes	var.	yes
Saebah	—	—	—	—	—
Chiangmai	yes	yes	yes	life	yes
Zanjera Danum Sitio	—	yes	yes	var.	yes
Tanowong B	yes	yes	yes	—	—
Pinagbayanan	yes	yes	yes	fixed	yes
Thulo Kulo	yes	yes	yes	var.	yes
Takkapala	—	—	—	—	—
Bondar Parhudagar	yes	yes	yes	fixed	yes
Nabagram	yes	yes	yes	—	—
Na Pae	yes	no	yes	fixed	yes
Cadchog	yes	no	yes	var.	no
Silean Banua	yes	yes	yes	fixed	yes

— = Missing in case.

* Denotes a case that is negative in either rule conformance or maintenance or both. Cases without asterisks are positive in both rule conformance and maintenance.

a. Fixed term, may be re-elected.

b. Variable term, subject to vote of confidence.

c. Life term.

the accounts of the organization are also presented and reviewed in the meetings. In both systems, other general meetings may be held throughout the year whenever major decisions concerning the operation of the system have to be made. General meetings are considered a major event in most of the community irrigation systems. In Oaig-Daya in the Philippines, farmers are even required to pay a fine for being absent from a general meeting.

Specialized officials or monitors are appointed to enforce operational rules in most of the community cases. In Calaoaan in the Philippines, for example, the chairman and the board members of the irrigators' association are responsible for organizing maintenance works. In Nabagram in Bangladesh, water is distributed successively from one block to another during the postplanting period. A water distributor is employed to determine when an individual plot has an adequate supply of water and to divert the water flow from one plot to another. With the water allocation process taken out of the hands of individual irrigators, the chance of rule violations is reduced. Provided that the water distributor is held accountable to irrigators, his service helps to reduce the chance of rule violations.

The chief executives in most of these collective-choice entities are selected through direct or indirect elections by appropriators. The periods for which the chief executives serve, however, vary from case to case. In some of the cases, officials are subject to periodic reelection. In Silean Banua, for example, the six officers on the board of directors are subject to reelection every two years. In other cases, officials can serve for an indefinite period of time, subject to a vote of nonconfidence by members. In Calaoaan in the Philippines, the members of the board of directors of the irrigators' association are selected by irrigators. Their term of office is unspecified. Their tenure ends only when they are incapable of exercising their duties or when they have committed major mistakes. A past chairman of the association, for example, had served for thirteen years and resigned from the post because of old age.

The chief executives are compensated in most of the cases. Some of the most common forms of compensation for irrigation officials in these cases include reduced labor obligations, reduced membership dues, and fines or direct payments, in the form of cash or agricultural products, by irrigators. In return for their services, the irrigation headmen in Chiangmai, for example, are excused from paying taxes on certain amounts of land, they do not have to contribute labor for maintenance, and they can keep some of the fines levied.[5]

There are, however, a few exceptions in which officials are not paid. In Diaz Ordaz Tramo, officials have to perform various duties including the organization of water allocation, maintenance, and conflict resolution. For these duties, the officials receive no compensation and little praise. Every landholder within the appropriation area, however, is obliged to occupy the positions through rotation; each has to take an office for one year. In Cadchog and Calaoaan in the Philippines, irrigation leaders are not compensated for their duties. Their own interests in the irrigation systems may have been a sufficient incentive for them to help govern the systems.

Bureaucratic systems.   In a bureaucratic irrigation system, the production resource is governed by national or regional government agencies. In some bureaucratic systems, the same agencies may govern the distribution and appropriation resources of a system. In others, different collective-choice entities, such as irrigators' associations, are involved in governing activities in the distribution or appropriation resources. In six of the bureaucratic cases in the sample, the appropriation resources are governed solely by government agencies. In the other eight cases, the appropriation resources are governed by both government agencies and local collective-choice entities constituted by appropriators.

Four questions about collective-choice rules in a government agency were asked in the coding form:

- *Financial source*: What is the major financial source of the agency?

- *Report to higher authority*: Do the administrators who make major operating decisions for the appropriation resource report to any external or higher-level authority?

- *Official near resource*: Do the administrators who make major operating decisions for the appropriation resource reside in or near the resource?

- *General meetings*: Can appropriators participate in general meetings to express their needs and concerns to the administrators who make major operating decisions for the appropriation resource?

The collective-choice rules used in most government agencies in the sample appear to be unfavorable to rule formulation, rule enforcement, or the accountability of officials to irrigators. As shown in Table 5.14, the major financial source of all these agencies, with the exception of Area Four in Taiwan, is government allocation. Because these officials and their agencies are not financially dependent on irrigators, they are usually not as motivated to serve irrigators as are their counterparts in irrigators' organizations. In all the cases, officials who are responsible for making major operating decisions concerning various appropriation resources are not irrigators themselves but full-time employees of government agencies. Instead of reporting to irrigators, they report to a higher authority within or outside their agencies.

The Provincial Irrigation Department that governs Gondalpur Watercourse in Pakistan, for example, receives funding for recurrent and operational expenditures through allocations by the Provincial Finance Department. The allocations are based on the physical

TABLE 5.14   Collective-Choice Rules in Government Agencies: Bureaucratic Irrigation Systems

	Financial source	Officials report to higher authority	Officials live near resource	General meetings held
Gondalpur Watercourse*	government	yes	no	no
Area One Watercourse*	government	yes	no	no
Area Two Watercourse*	government	yes	no	no
Dakh Branch Watercourse*	government	yes	no	no
Dhabi Minor Watercourse*	government	yes	no	no
Punjab Watercourse*	government	yes	no	no
Amphoe Choke Chai*	government	yes	yes	yes
Area Three Watercourse*	government	yes	yes	no
Kottapalle	government	yes	no	no
Nam Tan Watercourse	—	yes	yes	no
Sananeri	government	yes	no	no
Area Four Watercourse	irrigators	yes	yes	no
Kaset Samakee	government	yes	yes	yes
El Mujarilin	government	yes	yes	no

— = Missing in case.

* Denotes a case that is negative in either rule conformance or maintenance or both. Cases without asterisks are positive in both rule conformance and maintenance.

characteristics and inventory of the irrigation facilities. The Irrigation Department receives a fixed amount of funding per year for each kilometer of canal that exceeds a certain discharge capacity. The basis for budget allocations is rigidly fixed and often based on formulae that were established decades ago. The day-to-day field work of the department is carried out under the direction of the executive engineer at the divisional level, who is responsible for thousands of hectares of farmland. This official decides on the supply of water to various watercourses primarily on the basis of instructions from headquarters and the available water supply in the main river, not the conditions and demands in the command area. The Irrigation Department as a whole "can be fiscally accountable and fully responsible in [its] work and yet have minimal interaction with farmers, who often feel that the irrigation service they receive is not satisfactory" (Merrey and Wolf 1986, 10).

As shown in Table 5.14, in most of the cases, officials who make major decisions for watercourses reside far away from the appropriation resources they serve. These officials develop little identification with the interests of the local communities and have little incentive to be actively involved in solving farmers' problems. Their distance from the appropriation resources also prevents them from acquiring timely and accurate information about the various needs of the appropriation resources. In all but two cases, officials do not convene general meetings with irrigators. Irrigators themselves usually have few formal channels through which to articulate their interests and grievances to officials.

Robert Wade (1988a), in his study of Kottapalle in India, describes how the Irrigation Department's supervisor and assistant engineer relate to farmers there. These officials live and work far away from the village but control the level of water supply to the village. They may visit the village occasionally. During their visit, they will stay mostly with their local contact, a contractor in the village who works regularly for the Irrigation Department. Rather than helping farmers solve their irrigation problems, their purpose in visiting is usually to negotiate the bribe to be paid for assured water supplies to the village. When the water supply is exceptionally scarce, farmers in the village even have to send their representatives to the Irrigation Department in the town to get more water for the village through bribery. One cannot, of course, generalize from this case that every irrigation bureaucracy is corrupt. The case shows, however, that irrigation officials who are not properly motivated and monitored may create difficulties for farmers instead of helping them.

Complex bureaucratic irrigation systems governed solely by government agencies are unlikely to solve all water allocation and maintenance problems at the watercourse level. Within the sample, all six of the cases governed solely by government agencies are characterized by both a low degree of rule conformance and poor maintenance (Table 5.15). In these cases, operational rules handed down from government agencies often turn out to be incompatible with the special circumstances of individual watercourses.

In some of these bureaucratic irrigation systems, local farmers who are unable to develop their own collective-choice arrangements have to develop extralegal rules to suit their own circumstances. Merrey, for example, discusses the difference between informal,

TABLE 5.15   Rule Conformance and Maintenance by Local Collective-Choice Entities: Bureaucratic Irrigation Systems

	With local collective-choice entity	Without local collective-choice entity	Total
Cases rated positive in both rule conformance and maintenance	75% 6	0% 0	6
Cases rated negative in either rule conformance or maintenance or both	25% 2	100% 6	8
Total	100% 8	100% 6	14

Percentage difference = 75%
Chi-square with continuity correction factor = 5.1
D.F. = 1   P < 0.05

farmer-established rotations and formal rotations established by the Irrigation Department in Gondalpur Watercourse. He writes:

> Unlike the formal rotation, the informal rotation takes into consideration local conditions such as the sandiness of soils and the height of the field relative to the ditch. Thus, a sandy or high field is awarded extra time to ensure it can be irrigated. More time is also allowed for filling long sections of the watercourse. (Merrey and Wolf 1986, 46)

As discussed in Chapter 2, the effectiveness of operational rules depends on local circumstances. The involvement of cultivators in the formulation and enforcement of operational rules at the watercourse level facilitates the adaptation of these rules to the specific needs of different appropriation areas within a larger irrigation system. In some of the bureaucratic cases, local appropriators have constituted collective-choice entities that adopt and enforce their own operational rules at the watercourse level. Complex bureaucratic cases with local irrigators' organizations usually perform better than those without because operational rules developed and enforced by local collective-choice entities are usually more effective in meeting the needs of farmers. Among the bureaucratic cases in the sample, those with local collective-choice entities are characterized by a higher incidence of rule conformance and adequate maintenance than those without (see Table 5.15).

As shown in Table 5.16, local collective-choice rules in the bureaucratic cases are similar to the ones found in community irrigation systems. Most of the local collective-choice entities in the bureaucratic cases have general meetings that involve their members in major collective choices. Specialized officials or monitors are appointed by irrigators in most cases to oversee the implementation of operational rules within their appropriation resources. Executives, most of whom have fixed or variable terms of office, are selected by irrigators.

Despite these similarities, however, one should avoid making an unqualified analogy between irrigators' organizations in community irrigation systems and those within bureaucratic irrigation systems. Irrigators' organizations in community systems are self-contained entities, while those in bureaucratic systems are units within a larger organizational environment. Irrigators' organizations

TABLE 5.16   Collective-Choice Rules in Local Collective-Choice Entities: Bureaucratic Irrigation Systems

	General meetings held	Monitors to enforce rules	Elected executives	Tenure of executives	Paid executives
Amphoe Choke Chai*	yes	no	yes	fixed[a]	—
Area Three*	yes	yes	—	—	—
Kottapalle	yes	yes	yes	fixed	no
Nam Tan Watercourse	—	yes	yes	var.[b]	yes
Sananeri	yes	yes	yes	var.	no
Area Four Watercourse	yes	yes	yes	fixed	—
Kaset Samakee	yes	no	yes	fixed	—
El Mujarilin	no	—	no	life[c]	yes

— = Missing in case.

* Denotes a case that is negative in either rule conformance or maintenance or both. Cases without asterisks are positive in both rule conformance and maintenance.

a. Fixed term, may be re-elected.

b. Variable term, subject to vote of confidence.

c. Life term.

in bureaucratic systems will not be successful if irrigators fail to perceive a need to organize or if their organizations are unable to maintain sufficient autonomy in governing their own affairs.

Amphoe Choke Chai, for instance, is an irrigators' organization established under the auspices of the Royal Irrigation Department of Thailand to help govern two water zones within the Lam Pra Plerng Irrigation Project. Even with the encouragement of the government agency, the irrigators' organization has not been very successful in attracting members and organizing water allocation and maintenance because farmers can receive sufficient water from the natural flooding of rivers and therefore are not motivated to operate and maintain the canal networks that belong to the irrigation project.

Kottapalle provides an example in which farmers have been able to constitute their own collective-choice entity and enforce their own rules governing their investments and water allocation. The tasks of this entity are performed with neither support nor interference from the government agencies responsible for the irrigation system. Most government officials are not even aware of the existence of the entity. The real need for cooperation in water allocation and the lack of interference from government officials have enabled individual irrigators in Kottapalle to develop their own collective-choice arrangements to govern their appropriation resource.

In other cases, such as Nam Tan Watercourse and El Mujarilin, government officials and local leaders cooperate in governing a watercourse. In El Mujarilin, for example, an official representing the Ministry of the Interior is responsible for hearing complaints between irrigators. Unless the dispute involves a clear infraction of the civil code, however, it is common for the case to be referred to the leader of the local tribe or other tribesmen whom the petitioners might choose. This practice allows the traditional tribal organization to remain a viable instrument for solving conflicts among irrigators.

The Role of Physical and Community Attributes. Differences in collective-choice arrangements help to explain why the bureaucratic cases in the sample have a higher incidence of problems in rule conformance and maintenance than the community cases. It remains a puzzle, however, whether the bureaucratic and community cases differ in their performance only because bureaucratic systems tend to have less favorable physical and community features than community cases. As discussed in Chapter 4, inadequate supplies of water, major social cleavages, and high income variance among irrigators are attributes that tend to be associated with problems in rule conformance and maintenance. As shown in Table 5.17, the community cases in the sample are generally characterized by more favorable physical and community attributes than the bureaucratic cases. A majority of the community cases are characterized by an adequate supply of water, no major social cleavages, and low-to-moderate income variance. Twelve out of fourteen bureaucratic cases, on the other hand, are characterized by an inadequate supply of water. Many of them are also characterized by major social cleavages and high income variance. Within the fourteen bureaucratic cases, any one of the three attributes—major social cleavages, high income variance, or the absence of local collective-choice arrangements—is a sufficient condition for problems in rule conformance and maintenance.

Because of the limited size of the sample, it is not possible to determine whether institutional attributes or physical and community attributes are more important factors for explaining why the bureaucratic cases have a higher incidence of problems in rule conformance and maintenance than the community cases. One cannot begin to separate spurious effects from the actual effects of each of these attributes until a much larger set of cases is available.

Despite this limitation, several observations can be made about the configurations of attributes and outcomes shown in Table 5.17.

TABLE 5.17   Physical and Community Attributes and Collective-Choice Arrangements

	Adequacy[a]	Cleavages[b]	Income Variance[c]	Local collective-choice arrangements
COMMUNITY SYSTEMS				
Mauraro*	no	no	—	no
Chhahare Khola*	no	—	—	no
Naya Dhara*	no	no	—	no
Char Hazar*	no	no	—	yes
San Antonio 1*	no	no	—	yes
San Antonio 2*	no	no	—	yes
Oaig-Daya*	no	no	low	yes
Silag-Butir	no	no	—	yes
Tanowong T	no	no	—	yes
Sabangan Bato	no	no	—	yes
Nayband	no	no	low	no
Calaoaan	yes	no	—	yes
Felderin	yes	no	moderate	no
Kheri	yes	no	moderate	yes
Raj Kulo	yes	no	low	yes
Saebah	yes	no	—	yes
Chiangmai	yes	yes	high	yes
Zanjera Danum Sitio	yes	no	moderate	yes
Tanowong B	yes	no	—	yes
Pinagbayanan	yes	no	moderate	yes
Thulo Kulo	yes	no	low	yes
Takkapala	yes	no	—	yes
Deh Salm	yes	yes	moderate	no
Bondar Parhudagar	yes	no	low	yes
Nabagram	yes	no	moderate	yes

— = Missing in case.
* Denotes a case that is negative in either rule conformance or maintenance or both. Cases without asterisks are positive in both rule conformance and maintenance.
a. Adequacy of water supply.
b. Cultural or social differences among appropriators.
c. Variance in average annual family income among appropriators.

continued on next page

TABLE 5.17 *continued*

	Adequacy[a]	Cleavages[b]	Income variance[c]	Local collective-choice arrangements
COMMUNITY SYSTEMS				
Na Pae	yes	no	low	yes
Agcuyo	yes	no	—	no
Cadchog	yes	no	—	yes
Silean Banua	yes	no	—	yes
BUREAUCRATIC SYSTEMS				
Gondalpur Watercourse*	no	yes	high	no
Area One Watercourse*	no	—	high	no
Area Two Watercourse*	no	yes	high	no
Dakh Branch Watercourse*	no	yes	high	no
Dhabi Minor Watercourse*	no	yes	high	no
Punjab Watercourse*	no	yes	moderate	no
Amphoe Choke Chai*	no	no	—	yes
Area Three Watercourse*	no	no	moderate	yes
Kottapalle	no	no	moderate	yes
Nam Tan Watercourse	no	no	low	yes
Sananeri	no	no	moderate	yes
Area Four Watercourse	no	no	low	yes
Kaset Samakee	yes	no	—	yes
El Mujarilin	yes	no	moderate	yes
OTHER SYSTEMS				
Hanan Sayoc*	no	no	—	no
Lurin Sayoc 1*	no	no	—	no
Lurin Sayoc 2*	no	no	—	no
Diaz Ordaz Tramo	yes	—	low	yes

— = Missing in case.
* Denotes a case that is negative in either rule conformance or maintenance or both. Cases without asterisks are positive in both rule conformance and maintenance.
a. Adequacy of water supply.
b. Cultural or social differences among appropriators.
c. Variance in average annual family income among appropriators.

First, no single community case in the sample is simultaneously characterized by an inadequate supply of water, major social cleavages, and high income variance. One possible explanation is that farmers in such a situation would face so many obstacles for collective action that they would not have developed and sustained an irrigation system on their own in the first place. Farmers are capable of constructing and governing their own community irrigation systems mostly in situations that offer neither major physical nor community obstacles.

Second, unlike community irrigation systems, many bureaucratic irrigation systems are found in physical and social environments that are unfavorable to cooperation among farmers. One possible reason for this is that bureaucratic agencies are involved mostly in situations in which farmers have failed to develop their own community irrigation systems. Another possible explanation is that many bureaucratic irrigation systems are designed to deliver water to as many farmers as possible in order to justify their construction and maintenance costs. Furthermore, some bureaucratic irrigation systems in countries like India and Pakistan were developed as parts of large-scale settlement projects. In such settlement projects, individuals with divergent economic, social, and cultural backgrounds are brought together to share an irrigation system (see Merrey and Wolf 1986). Bureaucratic agencies in these situations are more likely to face serious governance problems than irrigators' organizations in community irrigation systems.

Third, although bureaucratic agencies are capable of constructing irrigation systems in areas characterized by unfavorable physical and community features, they may not be sufficient to ensure effective operation and maintenance of various appropriation areas. Six of the bureaucratic cases in the sample are characterized by two or three of the unfavorable physical and community features. All six cases are characterized by both a low degree of rule conformance and poor maintenance. Irrigators in these cases face a difficult situation: The bureaucratic agencies are not effective enough in enforcing rules and maintaining the appropriation resources, and irrigators are unable to develop local collective-choice arrangements to govern their own activities. Their situation can hardly be improved unless the governing capabilities of the bureaucratic agencies are improved considerably or special initiatives are undertaken to help irrigators develop their own local collective-choice arrangements.

Summary

Institutional arrangements in an irrigation system can be analyzed by reference to two levels—operational and collective-choice. Operational rules prescribe what must, may, or may not be done in relation to everyday operating activities in an irrigation system. To be effective for coordinating irrigators' activities, operational rules have to be compatible with the special circumstances of each irrigation system and appropriation resource. No single set of operational rules is good for all irrigation systems. A boundary requirement that includes more irrigators than the irrigation system can support will induce conflict among appropriators and make cooperation difficult. Allocation rules within an irrigation system have to be adjusted in light of changes in water supply. Labor inputs from appropriators can be mobilized on different bases depending on the amount needed to maintain a system. The effectiveness of a penalty rule depends on whether it is compatible with other institutional and community attributes.

Counterintentional consequences may arise if a uniform set of rules is imposed on a large area without consideration of local variations. One example concerns the use of land as the sole boundary requirement in many bureaucratic irrigation systems. Although the official policy in these systems is to ensure that everyone who needs water gets it, the actual outcome is that many farmers are unable to get the amount of water they are promised.

A greater diversity of operational rules is found among the community cases than the bureaucratic cases in the sample. In most irrigators' organizations, irrigators are regularly involved in major collective decisions for their own appropriation resources. In most bureaucratic systems, major decisions for an appropriation resource are made by officials who reside far away from the appropriation resource and have little identification with the irrigators. Irrigators' organizations are therefore more likely than government agencies to develop rules that suit their specific circumstances. Collective-choice rules adopted by most irrigators' organizations are also more conducive to rule enforcement and accountability of officials than those adopted in government agencies.

Differences in collective-choice arrangements help to explain why the community cases are characterized by higher incidences of rule conformance and good maintenance than the bureaucratic cases. The differences also help to explain why the bureaucratic cases with irrigators' organizations at the watercourse level are more likely to be characterized by a high degree of rule conformance and good maintenance.

Besides collective-choice arrangements, physical and community attributes also help to explain why the community cases are likely to perform better than the bureaucratic cases. Community irrigation systems are likely to be developed and sustained in situations with a reasonable supply of water, no major social cleavages, and low-to-moderate income variance among irrigators. A majority of the bureaucratic cases, on the other hand, are characterized by inadequate supplies of water. Many of them are also characterized by major social cleavages and high income variance. In comparison with irrigators' organizations in community irrigation systems, government agencies in bureaucratic irrigation systems are more likely to face serious collective-action problems.

CHAPTER SIX

Designing Complex Institutional Arrangements: Linking Bureaucratic and Local, Self-Governing Organizations

Many scholars and practitioners have recently begun to realize the limitation of relying primarily on government bureaucracies to solve development and collective-action problems (Esman and Uphoff 1984; Korten 1986, 1989; Uphoff 1986b; Werlin 1989). The failure of many large irrigation projects to deliver projected benefits to farmers illustrates this limitation of bureaucratic governance. Thus some scholars and practitioners have turned their attention to the importance of local, self-governing organizations in solving collective-action problems in irrigation systems.

Yet local indigenous organizations are not panaceas for all kinds of collective-action problems. Although it is incorrect to assume bureaucratic governance can solve all collective-action problems, it is equally wrong to assume that individuals at the local level can solve all the collective-action problems they face without drawing on external resources and facilities. Designers of irrigation projects in developing countries often assume that once water begins to flow in the main canal, farmers will spontaneously organize among themselves to build channels to divert water from the main canal to their fields. This assumption has, however, failed to materialize in many cases (Ascher and Healy 1990; Chambers 1988; Sengupta 1991).

In this study, I have illustrated the level of complexity involved in understanding factors that affect collective action in irrigation systems. On the one hand, the robustness of many local, self-governing irrigation organizations exemplifies some of the common arguments in favor of this type of organization. On the other hand, one can identify numerous obstacles to effective organization for collective action among farmers in irrigation systems. Some of these obstacles inhibit long-term cooperation among irrigators. Yet some of the obstacles can be overcome by the crafting of institutional arrangements that counteract perverse incentives inherent in various collective-action situations (see E. Ostrom 1992).

A complex, multilevel approach is needed to understand the terms and conditions for effectively organizing collective action in irrigation systems. I have drawn on and extended a theoretical framework derived from institutional analysis and transaction cost economics to examine how various physical, community, and institutional factors affect collective action in irrigation systems. Within the IAD framework, institutional arrangements are conceptualized as rules that structure repetitive, interdependent relationships among individuals. These constraints create stability of expectations among participants. If these rules are appropriately crafted and effectively enforced, they may create credible commitments among participants and thus facilitate their long-term cooperation.

In this concluding chapter, I first discuss several theoretical propositions derived from the present study and earlier work. Second, I examine the potential and limitations involved in utilizing local, self-governing organizations in governing irrigation systems. Finally, I argue that if irrigation development in developing countries is to be enhanced, it is necessary to reexamine the organization and policies of bureaucratic agencies and their relation to various local, self-governing organizations.

Propositions about Institutions and Collective Action

Several propositions about institutions and collective action can be derived from the present study and earlier work. First, *multiple layers of rules affect actions and outcomes in irrigation systems*. Operational rules define who can participate in which situations; what the participants may, must, or must not do; and how they will be rewarded or punished. Several types of operational rules—boundary, allocation, input, and penalty—can be used to coordinate water allocation and maintenance in irrigation systems. Another layer of rules—collective-choice rules—specifies the terms and conditions for adjudicating conflicts, enforcing decisions, and formulating operational rules. The effectiveness of a set of collective-choice arrangements depends on its ability (1) to help formulate operational rules that meet the needs of users, (2) to detect and provide sanctions against rule violations, and (3) to hold officials accountable to users.

Operational and collective-choice rules are nested in a hierarchy. Operational rules are enforced or changed within conditions stipulated by collective-choice rules. Operational rules can usually be changed more economically and more rapidly than collective-choice rules. The relative stability of collective-choice rules creates continuity for long-term cooperative arrangements.

Second, many types of collective-action problems are involved in irrigation. Each of these problems may involve a different community of individuals. *Multiple collective-choice entities are needed to solve collective-action problems of various scopes.* In many large-scale irrigation systems, for example, bureaucratic agencies play an important role in governing production and distribution facilities. At the appropriation level, farmers' organizations may develop and enforce operational rules governing water allocation and maintenance. Because more than one collective-choice entity is often involved in governing irrigation systems of considerable size, problems cannot be solved just by focusing on one organization. These different organizations are often unrelated to one another hierarchically. Each organization preserves a certain degree of autonomy. The challenge these organizations face is to develop mutually productive relationships among themselves. If participants see one another as the source of problems rather than as potential allies, it is difficult to develop productive interorganizational relationships.

Third, uncertainty is a major source of obstacles for human exchange and cooperation. Uncertainty often results from difficulties in measuring the respective contributions of participants to a common endeavor and in enforcing agreements among participants (North 1990). *Institutional arrangements that help to reduce uncertainty arising from within the group of participants and from other physical, social, or political circumstances motivate partici-pants to develop credible commitments and thus long-term coop-erative relationships with one another.* Creating an institutional framework that induces credible commitments among participants requires the crafting of a complex, interdependent set of opera-tional and collective-choice rules that are compatible with specific physical and social environments (see E. Ostrom 1992).

Fourth, *institution building is a long-term process that requires the investment of resources and extensive trial and error.* It often takes a long time and extensive efforts to develop and eventu-ally benefit from a set of functioning institutional arrangements. For example, to develop an appropriate set of operational rules requires careful experimentation and fine adjustments. Larger bu-reaucratic agencies may be able to develop uniform rules that deal with common problems shared by many farmers. Yet a great di-versity of institutional arrangements is needed to deal with var-ious context-specific problems. It is important to generate a va-riety of institutional arrangements to resolve the diverse prob-lems farmers encounter in different locations. Farmers must be actively involved in rule formulation and enforcement in order to organize water allocation and maintenance at the appropriation level.

Individuals have to overcome many obstacles before they can develop and benefit from a set of rules that effectively solve their common problems. They must, for example, secure information about the nature of the problem they face and the preferences of the individuals involved. They must also negotiate cooperative ar-rangements and devise ways to enforce them. In some cases, the costs of undertaking these activities are high enough to prevent participants from developing any resolution to the problem they face.

It takes time to develop and eventually benefit from a set of functioning institutional arrangements: Institution building is an in-vestment process whose returns can be realized only in the future. Individuals will invest in institution building only if they expect enough stability in the environment that their investments will not be wasted when circumstances change.

Fifth, *individuals will support an institutional change only if they believe the potential benefits to them of the new arrangement outweigh the potential costs.* If some individuals benefit dispro-portionately from the existing situation, they will resist changes that may decrease their benefits in the future. For example, head-enders usually have a more secure access to water, so they have less incentive to cooperate with tailenders in water allocation and maintenance. Institutional change is most feasible when everybody gains from the change. Robert Chambers writes, "Losers can be induced through the exercise of power to accept loss, but it is easier if they do not have to lose in the first place. Much of the search for a water revolution is a search therefore for ways in which head-enders can be better off and see themselves as better off with less water" (1988, 117).

In light of these propositions and the empirical evidence discussed in this and other studies, I next examine the potential and the limitations of local, self-governing organizations in the governance of irrigation systems.

The Potential and Limitations of Local, Self-Governing Organizations

The community irrigation systems examined in this study tend to be more effective in maintenance and water allocation than the bureaucratic systems. This does not, however, imply that community organizations are always superior to bureaucratic agencies in managing all types of irrigation problems. To understand the role of local, self-governing organizations in irrigation development, one must assess their potential and their limitations.

Local, self-governing organizations have several advantages over centralized bureaucratic agencies with respect to water allocation and maintenance. First, to be effective, water allocation and maintenance activities need to be organized in ways that are compatible with the specific social and physical environments. In appropriation areas that are subject to rapidly changing physical conditions, immediate actions are often needed to respond to various contingencies. As discussed earlier in this study, for example, allocation rules must be adjusted frequently in response to changing meteorological and cropping patterns. Even within one appropriation area, more than one set of allocation rules may be needed. In most cases, more restrictive allocation rules are used during times of water scarcity than during times of abundance. In some situations, community features affect the choice of allocation rules. In an appropriation resource with major social or cultural cleavages among irrigators, for example, conflicts can be reduced by arranging water schedules along major social or cultural divisions in the community.

In most situations, irrigators have the most intimate knowledge about their own physical and social environments. Furthermore, their proximity to the appropriation area enables them to utilize their knowledge effectively and to act quickly in solving problems. This is contrasted with many bureaucratic agencies in developing countries where the responsible officials live far away from the appropriation area and possess little knowledge about the physical and social environments at the local level.

Second, the shared community of understanding among participants affects the way they relate to one another. Rules can be effective only if a shared community of understanding exists among participants. A formal decree issued from a distant government agency will not be effective unless irrigators have a common understanding of the rules and have incentives to follow them. Rules developed within an indigenous community, on the other hand, often result from extended deliberation and experimentation by members of the community. After going through these processes, members of the community are more likely to have a common understanding and a common interest in the rules.

Third, individuals who have lived together for a sustained period may be able to develop various social networks and reciprocal relationships with one another. Although these networks and relationships may be established for other purposes, these same networks and relationships may facilitate cooperation in irrigation. Knowledge about these networks and relationships is shared by participants. It is often difficult for outsiders to ascertain this complex web of reciprocal relationships and networks. The participants themselves know better when and how they can utilize their social capital to undertake collective action.

Finally, rules adopted by farmers are also likely to be more relevant to local circumstances because farmers who decide to adopt the rules have to bear the consequences of their own decisions. Farmers' personal stakes in the irrigation system also motivate them to monitor one another's behavior. This is contrasted with officials in many bureaucratic agencies whose career advancement is unrelated to how well they serve farmers' interests.

Although indigenous organizations can play an important role in the governance of irrigation systems, these organizations are not always successful. The effective functioning of community organizations often requires a hospitable macropolitical regime. In countries where well-developed legal systems are available and the political system facilitates local public entrepreneurship, community organizations can be developed and sustained more easily than in other places. William Blomquist (1992), for example, documents how the courts and other aspects of the political system in California facilitated negotiated settlement among communities sharing groundwater basins in the area. In societies that do not recognize the rights of local communities to establish binding commitments within and among themselves, it is more difficult for local communities to solve their own collective-action problems.

In some situations, social and cultural divisions can inhibit coordination among irrigators. The costs of organizing collective action will be higher in communities that are divided by ethnic, caste, clan, or other cultural differences. In some situations, the divisions may be great enough to prevent any form of long-term cooperation among irrigators. In other situations, conflicts among different social groups are mitigated by organizing collective action within instead of across groups. This strategy, however, has its limitations, especially in situations that require cooperation involving large numbers of irrigators.

The range of activities that can be handled by community organizations may be limited. In some cases, indigenous institutional arrangements are tailored to suit certain physical and social environments. Even though a set of institutional arrangements may enable irrigators in an appropriation resource to solve their daily operation and maintenance problems, the arrangements may be ineffective under other circumstances. I discussed earlier in this study a number of community irrigation systems where appropriators have been able to develop and sustain operational rules that coordinate their water allocation and maintenance activities without explicit collective-choice arrangements. These irrigation systems can remain viable because they serve small numbers of irrigators or small irrigated areas and the water allocation and maintenance tasks in the systems are simple and straightforward. Irrigators in these systems may, however, face serious coordination problems when unexpected challenges arise.

Developing self-governing organizations is often costly, especially at the initial stage when participants are uncertain about the feasibility or consequences of alternative arrangements. It often requires extensive efforts to acquire information about the needs of individual participants and the effects of adopting different operational and collective-choice rules. If individuals are uncertain about the prospect of a cooperative arrangement or are unwilling to invest in the initial institution-building process, collective inaction and inefficiencies may be sustained for a long time.

Coordinating Bureaucratic Agencies to Local, Self-Governing Organizations

The preceding discussions show that local, self-governing organizations have both potential and limitations in their ability to solve collective-action problems in irrigation. Whether or not farmers can develop their self-governing potential is often affected by the design and policies of related government agencies.

Design Problems in Some Bureaucratic Irrigation Systems. The design of many bureaucratic irrigation systems often creates obstacles to local cooperation. Many bureaucratic irrigation systems, especially in India and Pakistan, are designed to deliver water to as many farmers and as extensive an area as possible in order to justify their construction and maintenance costs. Increasing the number of beneficiaries of irrigation projects is also a common way for politicians in these countries to gain electoral support (Repetto 1986). Irrigators who are entitled to withdraw more water than an irrigation system can support create serious collective-action problems that may undermine the long-term viability of the system. In irrigation systems characterized by a high degree of water scarcity, farmers who cultivate land only in the tail-end portion of a watercourse are usually disadvantaged compared to those in the head-end portion. In many cases, tailenders also happen to be from the poorer stratum of society. Inequity is sustained in these irrigation systems where the official policy is to spread water to as many farmers as possible. The difference between headenders and tailenders may create so much conflict that it is extremely difficult to sustain cooperative arrangements among them.

Furthermore, government agencies construct and govern irrigation systems mostly in locations where farmers have failed to develop their own community irrigation systems in the first place.

In some cases, bureaucratic irrigation systems are constructed and governed as part of a larger settlement project in which individuals of diverse economic and social backgrounds are brought together in one location. Their economic and social differences can hinder their cooperation in irrigation matters.

Because of these and other unfavorable factors, it is unrealistic to assume that irrigators would automatically develop effective institutional arrangements to govern their water allocation and maintenance activities whenever water begins to flow in the main canal of a bureaucratic irrigation system. Many large-scale irrigation projects in developing countries failed because of this false assumption. In their recent work, Ascher and Healy (1990) document the problems associated with the Jamua Irrigation Project in India. The project was designed to tap the Jamua River by constructing diversion works and extending the canal network on its left bank. The major objective was to provide irrigation to cultivators whose landholding varied from 0.5 to 10 hectares. The project was started in January 1965 and was scheduled for completion by May 1969.

Yet by 1974 only 31 percent of the target area had been brought under irrigation. The failure of the project authorities to provide field channels for 69 percent of the cultivable land was a major setback for the project. The project authorities assumed that once the canal system was constructed, farmers would willingly and jointly contribute their own labor to construct field channels to divert water from the canal to their field. The authorities failed to foresee, however, that farmers located near the canal would have little incentive to devote their efforts to constructing channels that would deliver water through their own fields into those of others. Ascher and Healy argue that the beneficiaries would likely agree to provide field channels cooperatively if that had been a precondition for initiating the project and its necessity had been explained to them beforehand. The technical experts who designed the project entirely neglected this social aspect of irrigation, and their negligence became a major source of problems for the project.

Enhancing Local, Self-Governing Organizations. The preceding examples suggest that before a government agency develops plans to involve irrigators in governing a bureaucratic irrigation system, it is important to consider irrigators' potential and limitations in self-governance. In many cases, government or development agencies can help irrigators overcome their limitations and realize their potential. This can be done not by imposing a single blueprint on irrigators but by facilitating interactions among irrigators.

There is no single blueprint for organizing irrigators in an appropriation resource. Although successful irrigators' organizations generally share similar kinds of collective-choice rules, the specifics of these rules differ from case to case. Leaders in most of these organizations, for example, are selected by irrigators and held accountable to irrigators; the specific arrangements by which these leaders are selected and compensated, however, differ from system to system. Cultivators in most of these organizations have to participate in general meetings that make major collective decisions for the organizations, but the specific formats of these meetings vary from case to case.

Some studies examine how government employees achieved impressive results by acting as catalysts in the irrigators' development of collective-choice arrangements. One example is an irrigation development project undertaken in Gal Oya, Sri Lanka (Uphoff 1985; E. Ostrom 1990). Organizers of the project explicitly rejected the idea of devising a single model of farmer organization for all 19,000 irrigators of the project. Instead, institutional organizers (IOs), mostly college graduates, were assigned to different appropriation areas in the project. Each IO had to live in the area and become familiar with the irrigators and the problems they faced. Instead of imposing a predefined organizational format on them, the IO tried to help irrigators organize special working committees to deal with specific problems they faced. After irrigators had experienced various working relationships with one another, the IO moved toward establishing an organization among irrigators and selecting leaders to represent the organization. This bottom-up method of organizing helped irrigators to develop their self-governing capabilities.

The way agencies relate to local collective-choice entities affects irrigators' incentives to participate in the governance of their appropriation areas. First, although government agencies can play an active role in settling disputes among different irrigators' organizations and providing legal guarantees for contractual arrangements, an irrigators' organization will not be effective unless it has considerable autonomy to decide on its internal matters. If agency officials can easily override the decisions of an irrigators' organization, irrigators will have little incentive to develop a consensus among themselves. Irrigators' organizations would have little chance to survive if irrigators have no right to mobilize their own resources and their operational and collective decisions are entirely dictated by agency officials.

Second, when an external authority attempts to provide assistance to a community irrigation system, there is a danger that the physical and community attributes of an appropriation resource may be changed. A new configuration of these attributes may, in turn, change the nature of collective-action problems in the resource, and new operational and collective-choice rules may be needed to deal with these problems. A development or government agency may, for example, attempt to help irrigators by rebuilding the headwork of their irrigation system or extending its distributory canals to more farmland. As a result, irrigators may encounter new collective-action problems. After new physical devices have been installed, new cooperative arrangements may be needed to operate and maintain them. After the distributory canals have been extended, more farmland will be involved and new water allocation and maintenance arrangements may be needed. Institutional arrangements that have been successful in the past may be insufficient to organize irrigators to undertake these new collective tasks.

In Nepal, for example, when external government funding for construction and maintenance was made available to a community irrigation system, farmers in the system soon developed an expectation that repairs and maintenance jobs would be done with government funding. Entrepreneurial energy in the irrigators' community was directed toward getting money or construction contracts from the government agency instead of organizing operation and maintenance among the irrigators themselves. Irrigators no longer undertook their own maintenance work, and conflict developed among them (see Fowler 1986).

Third, farmers are more likely to support their own irrigation organization if their participation is a precondition for receiving benefits from the system. A donor or government agency may, for example, require farmers of a community irrigation system to agree to repay a portion of the costs of a rehabilitation project. As long as the farmers have the right to decline an offer, this repayment requirement encourages them to develop binding commitments among themselves to ensure positive returns from the project (E. Ostrom, Schroeder, and Wynne 1990).

Fourth, farmers will have incentives to participate in a local organization only if they perceive that the benefits they obtain from the organization exceed the costs of the resources they devote to it. As discussed in this study, for example, if farmers have a highly unpredictable supply of water, which they believe that they have no effective means of affecting, they have little incentive to organize in the operation and maintenance of the appropriation area. Farmers are motivated to support their own organizations if these organizations are involved in the decision-making processes of irrigation projects. An irrigators' organization may participate in designing and supervising rehabilitation projects sponsored by government or development agencies. This kind of involvement not only helps to ensure that the rehabilitation project meets the needs of irrigators but can also help to strengthen the irrigators' organization.

Restructuring Government Agencies. The involvement of farmers in the governance of water allocation and maintenance activities is crucial in many large-scale irrigation systems, but such participation will be unsuccessful unless both the organizational problems of the bureaucratic machinery and the structures of incentives for farmers are corrected. Unfortunately, collective-choice rules in many bureaucratic agencies in developing countries are ineffective in formulating rules, enforcing rules, or holding officials accountable for their performance. These agencies are financially independent from irrigators because their major funding comes from government allocations. Officials in these agencies are usually not as motivated to serve their farmers as are their counterparts in irrigators' organizations. Officials who are responsible for making major operating decisions concerning various appropriation resources are not irrigators themselves but engineers who are professionally more interested in construction and engineering works than day-to-day operation and maintenance. Most of these officials do not expect to serve a particular area for a long time. Their career advancement usually depends on their formal qualifications and evaluations by their superiors rather than on how well they have served the irrigators.

In most cases, officials who make major decisions for watercourses reside in places far away. These officials develop little identification with the interests of local communities and have little incentive to be actively involved in solving farmers' problems. Their distance from the appropriation resource also prevents them from acquiring timely and accurate information about the various needs of different watercourses.

These institutional features create disincentives for officials to work in the best interests of farmers. In some cases, officials may even deliberately create uncertainties in water delivery to extract more bribes from farmers. The poor performance of irrigation officials makes farmers distrust them. From his own perspective, even if a farmer follows the allocation rule, he is not guaranteed the share of water to which he is entitled. After farmers have become accustomed to an anarchic form of water acquisition, officials can no longer count on them to follow rules and orders. This creates a vicious cycle. Officials and farmers do not trust one another and no institutional arrangement can be effective in disciplining water allocation and maintenance (Wade 1988b). Until the incentive structure within the government agency is corrected, farmers' participation alone may not significantly improve the overall performance of an irrigation system.

First, as discussed earlier, irrigation officials are not motivated to serve farmers' interests if their financial sources and careers are unrelated to their performance in operation and maintenance. A way to improve the performance of government agencies is to develop financial mechanisms that link the financial rewards of an agency with its ability to satisfy irrigators' needs. Studies have shown how this kind of arrangement promotes the performance of government agencies in operation and maintenance. The Philippine government, for example, initiated several changes at the National Irrigation Administration (NIA) in the 1970s. The budgetary grants for the operation and maintenance expenses of the NIA were gradually phased out, and the NIA was required to finance its operation and maintenance by collecting fees directly from farmers. This change induced officials in the NIA to attend to farmers' concerns and to promote the formation of water user associations that could help to collect water fees from farmers (Korten and Siy 1988).

Another example concerns the management accountability system in irrigation associations in Taiwan. According to Mick Moore, delay in payment of irrigation fees "has become an institutional mechanism, albeit an unpublicized one, through which farmers express dissatisfaction with the service they receive." Each level of the irrigation bureaucracy is required regularly to report its collection records to the upper level. At each level, delays in fee payment are construed as signs of farmers' dissatisfaction. Each unit's collection records affect its annual evaluation by its superior, which indirectly affects "salary increments, promotions, and access to additional resources." This system creates strong incentives for irrigation officials to serve farmers' interests (Moore 1989, 1743).

Second, the personal and professional interests of irrigation officials in operation and maintenance affect their incentives to attend to needs and conflicts among irrigators (Chambers 1988; Wade 1988b). In most bureaucratic agencies in south Asia, the engineers' professional interests are directed mostly toward construction and other major engineering works. These engineers usually have little interest in dealing with problems and conflicts among farmers who are located miles away from their offices in town. This is contrasted with irrigation associations in Taiwan where staff are mostly recruited from the local community. They spend their careers doing operation and maintenance works in the same location. This local affiliation helps to establish a sense of mutual concern and obligation between staff and farmers (Wade 1988b, 495).

Finally, the overall performance of an irrigation system depends on the development of mutually beneficial relationships between government and local, self-governing organizations. On the one hand, the performance and policies of government agencies at the system level affect irrigators' incentive to develop self-governing organizations at the appropriation level. On the other hand, local, self-governing organizations can support government agencies by helping collect fees and monitor the behavior of individual government units and officials. Building credible commitments between government officials and irrigators is key to improving bureaucratic irrigation systems.

Notes

Chapter 2

1. *Action situation* is a broader concept than *transaction* because the former deals with all kinds of social situations and the latter concerns specific exchanges between or among people.

2. Other outcomes such as cropping intensities and agricultural productivity are important too. I do not focus on them in this study because information about them is generally absent in case studies on irrigation systems. For a more extensive list of "objectives in irrigation management," see Uphoff 1986a, 20–21.

3. This is only a partial list of physical and community attributes that may affect collective action in irrigation systems. Other attributes such as meteorological conditions and irrigators' cosmological views are also important variables to consider. They are not discussed in this study because most cases do not provide detailed information about them.

4. In some situations, water abundance may be a problem by itself and requires collective action for its solution. For example, if the water flow is so abundant as to create drainage problems or threaten the physical integrity of the water diversion or delivery works, farmers may be induced to undertake intensive collective efforts to keep their system in working condition.

5. Wittfogel distinguishes between two types of irrigated agriculture — hydroagriculture and hydraulic agriculture. Hydroagriculture refers to small-scale agriculture for which "strictly local tasks of digging, damming, and water distribution can be performed by a single husbandman, a single family, or a small group of neighbors, and in this case no far-reaching organizational steps are necessary" (Wittfogel 1981, 18). Hydraulic agriculture, on the other hand, deals with large amounts of water and requires elaborate organizational discipline to work.

6. Researchers have, however, discovered through experimentation that rice does not require a continuous stand of water during the growth period but that continuous flow of water through the field is necessary. If farmers could follow a rotational schedule for distributing water, a larger area could be cultivated by the same amount of water (Abel 1977).

7. There are, of course, other kinds of operational rules. I do not focus on them because they are in general less important than the four I am discussing here. I will discuss other operational rules whenever relevant.

8. Individuals who control a water resource may, in certain circumstances, use their control to increase the price of the crops they raise by means of that resource. This situation, however, would happen only in isolated communities that have no connection with other marketing networks. In locations that have regular connections with other marketing networks, individuals monopolizing a water source could not have much influence on the price of the crops they raise.

9. In the irrigation literature a distinction is often made between *water allocation* and *water distribution*. Martin and Yoder, for example, argue that

> water allocation is the assignment of entitlement to water from a system, both identifying the fields and farmers with access to water from the system and the amount and timing of the water to be delivered to each. Water distribution refers to the physical delivery of water to the fields and may not conform to the water allocation. (1986, 2)

Water allocation, as defined by Martin and Yoder, is analogous to what are called *bases* in this study; *water distribution* is analogous to water delivery *procedures*.

10. In larger irrigation systems, three or even more levels of collective-choice entities may exist.

Chapter 3

1. There is, however, no easy solution to the problem of defining the boundaries of an irrigation system. Hunt, for example, argued at one point that an irrigation system refers to the area receiving water from a single point on a natural water source. He later admitted to the limitation of this definition because some irrigation systems receive water from more than one source (Hunt 1978).

2. There is potentially a fifth stage, drainage. Because most of the case studies do not contain information about drainage activities and arrangements, these problems are not discussed in this study.

3. In some irrigation systems, especially those in Africa, government or parastatal bodies are responsible for both irrigation and cropping patterns in state farms (see Thornton 1976, 149). In these irrigation systems, the *use resource* is an integral part of the systems. In most of the cases examined in this study, use resources belong to private individuals. I do not discuss this particular resource in this study unless it is related to some other resources or collective activities within an irrigation system.

4. This distinction between simple and complex irrigation systems is different from that suggested by Spooner (1974), who uses the level of technological sophistication to distinguish simple irrigation systems from complex ones.

5. Collective-action problems occur at the production and distribution stages too. In some large-scale, complex irrigation systems, the problems may even involve multiple communities and government jurisdictions. An analysis of collective-action problems at these two stages requires even more complicated research design.

6. Chambers (1977) calls the latter a *bureaucratic-communal irrigation system*.

7. Another part of the research project is to set up laboratory experiments that resemble typical action situations faced by participants in common-pool resources. Researchers then examine how participants in an experiment respond to different structures of incentives induced by them (see E. Ostrom and Walker 1991; E Ostrom, Walker, and Gardner 1990).

8. For a discussion of the strengths and weaknesses of the case survey method, see Yin and Heald 1975.

Chapter 4

1. Another outcome—the distribution of benefits and costs among cultivators—will be discussed later in this chapter.

2. In the original coding form, the variable for the level of water supply has five values: (1) extreme shortage, (2) moderate shortage, (3) apparent balance, (4) moderate abundance, and (4) significant abundance. The following table shows how these five values are related to rule conformance and maintenance. Since no case has been coded for "significant abundance," it is not possible to examine the argument, mentioned in Chapter 2, that there will be little collective action by farmers when the supply of water is abundant.

TABLE N.1   Rule Conformance and Maintenance by Adequacy of Water Supply (in five values)

	Extreme shortage	Moderate shortage	Apparent balance	Moderate abundance	Significant abundance	Total
Cases rated positive in both rule conformance and maintenance	65% 2	32% 6	100% 19	100% 2	0	29
Cases rated negative in either rule conformance or maintenance or both	35% 5	68% 13	0% 0	0% 0	0	18
Total	100% 7	100% 19	100% 19	100% 2	0	47

3. The two systems, however, differ in terms of institutional arrangements (see Table 5.17 in Chapter 5).

Chapter 5

1. The constitutional-choice level that pertains to choices in collective-choice rules also has profound effects on any long-term cooperative efforts among humans (see V. Ostrom 1987). Unfortunately, I am unable to address problems related to constitutional-choice rules because they are rarely discussed in the case studies.

2. Of the four cases listed under "Other Systems" in Table 5.1, Lurin Sayoc 1 is governed by barriowide rural political officials. The other three are governed by municipal governments.

3. Personal communication.

4. Where there is an abundant supply of water, however, allocation rules may not be necessary. One example is an irrigation system in the Philippines, Nazareno-Gamutan, where water is so abundant that appropriators can have a continuous supply and no allocation rule is needed (see Ongkingko 1973).

5. Letting officials keep the fines they levy is a way to motivate them to monitor and impose sanctions on rule breakers. The arrangement may also have a perverse effect of encouraging officials to impose fines on irrigators indiscriminately. This perverse effect, however, is counterbalanced by other arrangements such as elections and general meetings that hold officials accountable to irrigators.

References

Abel, Martin G. 1977. "Irrigation Systems in Taiwan: Management of a Decentralized Public Enterprise." In *Water Resources Problems in Developing Countries*, Bulletin No. 3, ed. K. William Easter and Lee R. Martin. Minneapolis: University of Minnesota, Economic Development Center.

Anderson, Terry L. 1983. *Water Crisis: Ending the Policy Drought*. Washington, D.C.: Cato Institute.

Ascher, William, and Robert Healy. 1990. *Natural Resource Policy Making in Developing Countries*. Durham, N.C.: Duke University Press.

Bacdayan, Albert S. 1980. "Mountain Irrigators in the Philippines." In *Irrigation and Agricultural Development in Asia*, ed. E. Walter Coward. Ithaca, N.Y.: Cornell University Press.

Blomquist, William. 1992. *They Prefer Chaos*. San Francisco: ICS Press. Forthcoming.

Bottrall, Anthony. 1981. "Comparative Study of the Management and Organization of Irrigation Projects." World Bank Working Paper No. 458. Washington, D.C.: World Bank.

Bromley, Daniel W. 1982. "Improving Irrigated Agriculture: Institutional Reform and the Small Farmer." World Bank Working Paper No. 531. Washington, D.C.: World Bank.

———. 1984. "Property Rights and Economic Incentives for Resource and Environmental Systems." Agricultural Economics Staff Paper Series No. 231. Madison: University of Wisconsin.

Buchanan, James, and Gordon Tullock. 1962. *The Calculus of Consent: Logical Foundations of Constitutional Democracy*. Ann Arbor: University of Michigan Press.

Chambers, Robert. 1977. "Men and Water: The Organization and Operation of Irrigation." In *Green Revolution? Technology and Change in Rice-growing Areas of Tamil Nadu and Sri Lanka*, ed. B. H. Farmer. Boulder, Colo.: Westview Press.

———. 1988. *Managing Canal Irrigation: Practical Analysis from South Asia*. New York: Cambridge University Press.

Coward, E. Walter, Jr. 1979. "Principles of Social Organization in an Indigenous Irrigation System." *Human Organization* 38, no. 1: 28–36.

———. 1980a. "Management Themes in Community Irrigation Systems." In *Irrigation and Agricultural Development in Asia*, ed. E. Walter Coward, Ithaca, N.Y.: Cornell University Press.

———. 1980b. "Local Organization and Bureaucracy in a Lao Irrigation Project." In *Irrigation and Agricultural Development in Asia*, ed. E. Walter Coward. Ithaca, N.Y.: Cornell University Press.

———. 1986. "Direct or Indirect Alternatives for Irrigation Investment and the Creation of Property." In *Irrigation Investment, Technology, and Management Strategies for Development*, ed. K. William Easter. Boulder, Colo.: Westview Press.

Coward, E. Walter, Jr., and Ahmed Badaruddin. 1979. "Village, Technology, and Bureaucracy: Patterns of Irrigation in Comilla District, Bangladesh." *Journal of Developing Areas* 31: 431–40.

Cruz, Federico A. 1975. "The Pinagbayanan Farmers' Association and Its Operation." In *Water Management in Philippine Irrigation Systems: Research and Operations*, ed. International Rice Research Institute. Los Banos, Philippines: International Rice Research Institute.

Decentralization: Finance and Management Project. 1988. "Initial Research Agenda: Abstracted from Proposal for Decentralization Program Support in Nepal."

de los Reyes, Romana P., S. Borlavian, G. Gatdula, and M. F. Viado. 1980b. "Communal Gravity Systems: Four Case Studies." Quezon City, Philippines: Ateneo de Manila University, Institute of Philippine Culture.

de los Reyes, Romana P., et al. 1980a. "Forty-seven Communal Gravity Systems: Organization Profiles." Quezon City, Philippines: Ateneo de Manila University, Institute of Philippine Culture.

Downing, Theodore. 1974. "Irrigation and Moisture-Sensitive Periods: A Zapotec Case." In *Irrigation's Impact on Society*, ed. Theodore Downing and McGuire Gibson. Tucson: University of Arizona Press.

Esman, Milton J., and Norman T. Uphoff. 1984. *Local Organizations: Intermediaries in Rural Development*. Ithaca, N.Y.: Cornell University Press.

Fernea, Robert A. 1970. *Shaykh and Effendi: Changing Patterns of Authority among the El Shabana of Southern Iraq*. Cambridge, Mass.: Harvard University Press.

Field, Barry C. 1986. "Induced Changes in Property-Rights Institutions." Research Paper Series #86-1. Amherst: University of Massachusetts, Department of Agricultural and Resource Economics.

Fowler, Darlene, ed. 1986. "Rapid Appraisal of Nepal Irrigation Systems." WMS Report 43. Fort Collins: Colorado State University, Water Management Synthesis Project.

Gardner, Roy, Elinor Ostrom, and James Walker. 1990. "The Nature of Common-Pool Resource Problems." *Rationality and Society* 2, no. 3: 335–58.

Geertz, Clifford. 1980. "Organization of the Balinese Subak." In *Irrigation and Agricultural Development in Asia*, ed. E. Walter Coward. Ithaca, N.Y.: Cornell University Press.

Gillespie, Victor A. 1975. "Farmer Irrigation Associations and Farmer Cooperation." East-West Food Institute Paper No. 3. Honolulu: Food Institute, East-West Center.

Glick, Thomas F. 1970. *Irrigation and Society in Medieval Valencia.* Cambridge, Mass.: Harvard University Press.

Gray, Robert. 1963. *The Sonjo of Tanganyika.* Oxford: Oxford University Press.

Gustafson, W. E., and R. B. Reidinger. 1971. "Delivery of Canal Water in North India and West Pakistan." *Economic and Political Weekly* 6: A157–A162.

Hafid, Anwar, and Yujiro Hayami. 1979. "Mobilizing Local Resources for Irrigation Development: The Subsidi Desa Case of Indonesia." In *Irrigation Policy and the Management of Irrigation Systems in Southeast Asia*, ed. Donald C. Taylor. Bangkok: Agricultural Development Council.

Hardin, Garrett. 1968. "The Tragedy of the Commons." *Science* 162 (December): 1243–48.

Harriss, John. 1977. "Problems of Water Management in Hambantota District." In *Green Revolution? Technology and Change in Rice-growing Areas of Tamil Nadu and Sri Lanka*, ed. B. H. Farmer. Boulder, Colo.: Westview Press.

Hayami, Yujiro, and Vernon W. Ruttan. 1985. *Agricultural Development: An International Perspective.* Rev. ed. Baltimore, Md.: Johns Hopkins University Press.

Hayek, F. A. 1948. *Individualism and Economic Order.* Chicago: University of Chicago Press.

Hechter, Michael. 1987. *Principles of Group Solidarity.* Berkeley: University of California Press.

Hunt, Robert. 1978. "The Local Social Organization of Irrigation Systems: Policy Implications of its Relationship to Production and Distribution." Paper prepared under USAID contract.

———. 1989. "Appropriate Social Organization? Water User Associations in Bureaucratic Canal Irrigation Systems." *Human Organization* 48, no. 1: 79–90.

Joskow, Paul L. 1988. "Asset Specificity and the Structure of Vertical Relationships: Empirical Evidence." *Law, Economics, & Organization* 4, no. 1: 95–118.

Kiser, Larry L., and Elinor Ostrom. 1982. "The Three Worlds of Action. A Metatheoretical Synthesis of Institutional Approaches." In *Strategies of Political Inquiry*, ed. Elinor Ostrom. Beverly Hills, Calif.: Sage.

Korten, David C., ed. 1986. *Community Management: Asian Experience and Perspectives*. West Hartford, Conn.: Kumarian Press.

————. 1989. "The Community: Master or Client? A Reply." *Public Administration and Development* 9: 569–75.

Korten, Frances F., and Robert Y. Siy. 1988. *Transforming a Bureaucracy: The Experience of the Philippine National Irrigation Administration*. West Hartford, Conn.: Kumarian Press.

Lando, Richard P. 1979. "The Gift of Land: Irrigation and Social Structure in a Toba Village." Ph.D. diss., University of California, Riverside.

Langlois, Richard N., ed. 1986. *Economics as a Process*. New York: Cambridge University Press.

Lowdermilk, Max K., Wayne Clyma, and Alan C. Early. 1975. "Physical and Socio-Economic Dynamics of a Watercourse in Pakistan's Punjab: System Constraints and Farmers' Responses." Water Management Technical Report No. 42. Fort Collins: Colorado State University.

Maass, Arthur, and Raymond L. Anderson. 1986. *. . . and the Desert Shall Rejoice: Conflict, Growth, and Justice in Arid Environments*. Malabar, Fla.: Krieger.

Martin, Edward, and Robert Yoder. 1983a. "Water Allocation and Resource Mobilization for Irrigation: A Comparison of Two Systems in Nepal." Paper presented at the annual meeting of the Nepal Studies Association, Nov. 4–6, University of Wisconsin, Madison.

————. 1983b. "The Chherlung Thulo Kulo: A Case Study of a Farmer-Managed Irrigation System." In *Water Management in Nepal*, appendix 1, 203–17. Kathmandu: Ministry of Agriculture.

————. 1986. "Institutions for Irrigation Management in Farmer-Managed Systems: Examples from the Hills of Nepal." Research Paper No. 5. Digana Village, Sri Lanka: International Irrigation Management Institute.

Martin, Fenton. 1989. *Common Pool Resources and Collective Action: A Bibliography*. Bloomington: Indiana University, Workshop in Political Theory and Policy Analysis.

Meinzen-Dick, Ruth S. 1984. "Local Management of Tank Irrigation in South India: Organization and Operation." Cornell Studies in Irrigation No. 3. Ithaca, N.Y.: Cornell University.

Merrey, Douglas J., and James M. Wolf. 1986. "Irrigation Management in Pakistan: Four Papers." Research Paper No. 4. Digana Village, Sri Lanka: International Irrigation Management Institute.

Mirza, A. H. 1975. "A Study of Village Organizational Factors Affecting Water Management Decision Making in Pakistan." Water Management Technical Report No. 34. Fort Collins: Colorado State University.

Mirza, A. H., and Douglas J. Merrey. 1979. "Organization Problems and Their Consequences on Improved Watercourses in Punjab." Fort Collins: Colorado State University, Water Management Research Project.

Mitchell, William P. 1976. "Irrigation and Community in the Central Peruvian Highlands." *American Anthropologist* 78: 25–44.

———. 1977. "Irrigation Farming in the Andes: Evolutionary Implications." In *Studies in Peasant Livelihood*, ed. Rhoda Halperin and James Dow. New York: St. Martin's Press.

Moore, Mick. 1989. "The Fruits and Fallacies of Neoliberalism: The Case of Irrigation Policy." *World Development* 17, no. 11: 1733–50.

Nachmias, David, and Chava Nachmias. 1987. *Research Methods in the Social Sciences*. 3d ed. New York: St. Martin's Press.

Netting, Robert McC. 1974. "The System Nobody Knows: Village Irrigation in the Swiss Alps." In *Irrigation's Impact on Society*, ed. Theodore Downing and McGuire Gibson. Tucson: University of Arizona Press.

———. 1981. *Balancing on an Alp: Ecological Change and Continuity in a Swiss Mountain Community*. New York: Cambridge University Press.

North, Douglass C. 1990. *Institutions, Institutional Change and Economic Performance*. New York: Cambridge University Press.

Oakerson, Ronald J. 1986. "A Model for the Analysis of Common Property Problems." In *Proceedings of the Conference on Common Property Resource Management*, ed. National Research Council. Washington, D.C.: National Academy Press.

Olson, Mancur. 1965. *The Logic of Collective Action, Public Goods and the Theory of Groups*. Cambridge, Mass.: Harvard University Press.

Ongkingko, Petronio S. 1973. "Case Studies of Laoag-Vintar and Nazareno-Gamutan Irrigation System." *Philippine Agriculturist* 59, nos. 9 and 10: 374–80.

Ostrom, Elinor. 1986. "An Agenda for the Study of Institutions." *Public Choice* 48: 3–25.

———. 1988. "Institutional Arrangements and the Commons Dilemma." In *Rethinking Institutional Analysis and Development,* ed. Vincent Ostrom, David Feeny, and Hartmut Picht, 101–39. San Francisco: ICS Press.

———. 1990. *Governing the Commons: The Evolution of Institutions for Collective Action*. New York: Cambridge University Press.

———. 1992. *Crafting Institutions for Self-Governing Irrigation Systems*. San Francisco: ICS Press. Forthcoming.

Ostrom, Elinor, Larry Schroeder, and Susan Wynne. 1990. *Institutional Incentives and Rural Infrastructure Sustainability.* Burlington, Vt.: Associates in Rural Development, Inc.

Ostrom, Elinor, and James Walker. 1991. "Communication in a Commons: Cooperation without External Enforcement." In *Laboratory Research in Political Economy*, ed. Thomas R. Palfrey, 287–322. Ann Arbor: University of Michigan Press.

Ostrom, Elinor, James Walker, and Roy Gardner. 1990. "Sanctioning by Participants in Collective Action Problems." Paper presented at the Conference on Experimental Research on the Provision of Public Goods and Common Pool Resources, Indiana University, Bloomington, May 18–20.

Ostrom, Vincent. 1987. *The Political Theory of a Compound Republic: Designing the American Experiment.* 2d ed. Lincoln: University of Nebraska Press.

———. 1989. *The Intellectual Crisis in American Public Administration.* 2d ed. Tuscaloosa: The University of Alabama Press.

Ostrom, Vincent, and Elinor Ostrom. 1977. "Public Goods and Public Choices." In *Alternatives for Delivering Public Services: Toward Improved Performance*, ed. E. S. Savas, 7–49. Boulder, Colo.: Westview Press.

Palanisami, K. 1982. "Managing Tank Irrigation Systems: Basic Issues and Implications for Improvement." Paper presented at a Workshop on Tank Irrigation: Problems and Prospects.

Palanisami, K., and K. William Easter. 1986. "Management, Production, and Rehabilitation in South Indian Irrigation Tanks." In *Irrigation Investment, Technology, and Management Strategies for Development*, ed. K. William Easter. Boulder, Colo.: Westview Press.

Plott, Charles, and Robert A. Meyer. 1975. "The Technology of Public Goods, Externalities, and the Exclusion Principle." In *Economic Analysis of Environmental Problems*, ed. Edwin S. Mills. New York: National Bureau of Economic Research.

Postel, Sandra. 1990. "Saving Water for Agriculture." In *State of the World 1990: A Worldwatch Institute Report on Progress toward a Sustainable Society*, ed. Lester Brown et al. New York: Norton.

Potter, Jack M. 1976. *Thai Peasant Social Structure.* Chicago: University of Chicago Press.

Pradhan, Prachanda. 1983. "Chhatis Muaja." In *Water Management in Nepal: Proceedings of a Seminar on Water Management Issues.* Kathmandu, Nepal: Agricultural Projects Service Centre.

———. 1988. "Increasing Agricultural Production in Nepal: Role of Low Cost Irrigation Development through Farmer Participation." Kathmandu, Nepal: International Irrigation Management Institute.

Putterman, Louis, ed. 1986. *The Economic Nature of the Firm.* New York: Cambridge University Press.

Reidinger, Richard B. 1974. "Institutional Rationing of Canal Water in Northern India: Conflict between Traditional Patterns and Modern Needs." *Economic Development and Cultural Change* 23 (October): 79–104.

———. 1980. "Water Management by Administrative Procedures in an Indian Irrigation System." In *Irrigation and Agricultural Development in Asia*, ed. E. Walter Coward. Ithaca, N.Y.: Cornell University Press.

Repetto, Robert. 1986. *Skimming the Water: Rent-seeking and the Performance of Public Irrigation Systems.* Research Report No. 4. Washington, D.C.: World Resources Institute.

Schlager, Edella, and Elinor Ostrom. 1992. "Property-Rights Regimes and Natural Resources: A Conceptual Analysis." *Land Economics.* Forthcoming.

Sengupta, Nirmal. 1991. *Managing Common Property: Irrigation in India and the Philippines.* Newbury Park, Calif.: Sage.

Simon, Herbert A. 1961. *Administrative Behavior.* 2d ed. New York: Macmillan.

Siy, Robert Y., Jr. 1982. *Community Resource Management: Lessons from the Zanjera.* Quezon City, Philippines: University of the Philippines Press.

Spooner, Brian. 1971. "Continuity and Change in Rural Iran: The Eastern Deserts." In *Iran: Continuity and Variety*, ed. Peter J. Chelkowski. New York: New York University, Center for Near Eastern Studies and the Center for International Studies.

———. 1972. "The Iranian Deserts." In *Population Growth: Anthropological Implications*, ed. Brian Spooner. Cambridge, Mass.: MIT Press.

———. 1974. "Irrigation and Society: The Iranian Plateau." In *Irrigation's Impact on Society*, ed. Theodore Downing and McGuire Gibson. Tucson: University of Arizona Press.

Tang, Shui Yan. 1989. "Institutions and Collective Action in Irrigation Systems." Ph.D. diss., Department of Political Science, Indiana University, Bloomington.

———. 1991. "Institutional Arrangements and the Management of Common-Pool Resources." *Public Administration Review* 51, no. 1: 42–51.

Tan-kim-yong, Uraivan. 1983. "Resource Mobilization in Traditional Irrigation Systems of Northern Thailand: A Comparison between the Lowland and the Upland Irrigation Communities." Ph.D. diss., Cornell University.

Thornton, D. S. 1976. "The Organisation of Irrigated Areas." In *Policy and Practice in Rural Development*, ed. Guy Hunter, A. H. Bunting, and Anthony Bottrall. London: Croom Helm.

Uphoff, Norman. 1985. "People's Participation in Water Management: Gal Oya, Sri Lanka." In *Public Participation in Development Planning and Management*, ed. Jean-Claude Garcia-Zamor. Boulder, Colo.: Westview Press.

———. 1986a. *Improving International Irrigation Management with Farmer Participation: Getting the Process Right*. Boulder, Colo.: Westview Press.

———. 1986b. *Local Institutional Development: An Analytical Sourcebook with Cases*. West Hartford, Conn.: Kumarian Press.

Uphoff, Norman, M. L. Wickramasinghe, and C. M. Wijayaratna. 1990. " 'Optimum' Participation in Irrigation Management: Issues and Evidence from Sri Lanka." *Human Organization* 29, no. 1: 26–40.

Vander Velde, Edward J. 1971. "The Distribution of Irrigation Benefits: A Study in Haryana, India." Ph.D. diss., The University of Michigan.

———. 1980. "Local Consequences of a Large-Scale Irrigation System in India." In *Irrigation and Agricultural Development in Asia*, ed. E. Walter Coward. Ithaca, N.Y.: Cornell University Press.

Wade, Robert. 1984. "On the Sociology of Irrigation: How Do We Know the Truth about Canal Performance?" *Agricultural Administration* 19: 63–79.

———. 1985. "Common Property Resource Management in South Indian Villages." Paper prepared for a Conference on the Management of Common Property Resources in the Third World organized by the National Research Council.

———. 1988a. *Village Republics: Economic Conditions for Collective Action in South India*. New York: Cambridge University Press.

———. 1988b. "The Management of Irrigation Systems: How to Evoke Trust and Avoid Prisoner's Dilemma." *World Development* 16, no. 4.: 489–500.

Water and Engineering Commission. 1987. *Rapid Appraisal Study of Eight Selected Micro-Areas of Farmers' Irrigation Systems*. Final Report. Kathmandu, Nepal: Ministry of Water Resources.

Weissing, Franz, and Elinor Ostrom. 1991. "Irrigation Institutions and the Games Irrigators Play." In *Game Equilibrium Models*. Vol. 2, *Methods, Morals, and Markets*, ed. R. Selten, 188–262. Berlin: Springer-Verlag.

Werlin, Herbert. 1989. "The Community: Master or Client?—A Review of the Literature." *Public Administration and Development* 9: 447–57.

Wickham, T. H., and A. Valera. 1979. "Practices and Accountability for Better Water Management." In *Irrigation Policy and the Management of Irrigation Systems in Southeast Asia*, ed. D. C. Taylor and T. H. Wickham. Bangkok: Agricultural Development Council.

Williamson, Oliver. 1975. *Markets and Hierarchies*. New York: Free Press.

———. 1985. *The Economic Institutions of Capitalism*. New York: Free Press.

Wittfogel, Karl A. 1981. *Oriental Despotism: A Comparative Study of Total Power*. New York: Vintage Books.

Yin, Robert K., and Karen A. Heald. 1975. "Using the Case Survey Method to Analyze Policy Studies." *Administrative Science Quarterly* 20: 371–81.

ORDER FORM

Please accept this order for the following book:

	Qty	Price	Total
Tang, **Institutions and Collective Action:** **Self-Governance in Irrigation** (paper)	_____	x $9.95 =	_____
Ostrom, **Crafting Institutions for Self-** **Governing Irrigation Systems** (paper)	_____	x $9.95 =	_____

Subtotal _____

Shipping charges _____

CA residents please add applicable state and local sales tax _____

TOTAL _____

Shipping Charges: In North America, $3.00 for first book, $.75 for each additional book. Outside North America, $3.00 per book surface mail, $10.00 per book air mail.

Please include payment with order or provide full credit card information. Personal checks are accepted when drawn in U.S. funds on a U.S. bank.

Name _____

Institution _____

Address _____

City _____ State _____ Zip _____

Country _____

☐ Check enclosed ☐ MasterCard ☐ VISA

Credit Card # _____

Expires _____ Signature _____

☐ Please add my name to the ICS Press mailing list.

MAIL orders to: **ICS Press, 243 Kearny St., San Francisco, CA 94108**
FAX orders to : **(415) 986-4878**
PHONE orders to: **(800) 326-0263 toll free** in the U.S., or **(415) 981-5353**

Quantity discounts are available; please call (800) 326-0263 for details.